Lecture Notes in Mathematics 1483

Eduard Reithmeier

Periodic Solutions of Nonlinear Dynamical Systems

Numerical Computation, Stability, Bifurcation and Transition to Chaos

Springer-Verlag
Berlin Heidelberg New York
London Paris Tokyo
Hong Kong Barcelona
Budapest

Author

Eduard Reithmeier
5134 Etcheverry Hall
Department of Mechanical Engineering
University of California, Berkeley
Berkeley, CA 94720, USA

Cover: Leupold's draft for periodic perpetual motion. *From:* Jacob Leupold. Theatrum Machinarum Generale. Schauplatz des Grundes mechanischer Wissenschaften. Mathematico und Mechanico der königlich Preußischen und Sächsischen Societät der Wissenschaften, vol 1, Leipzig, Zunkel, 1724.

Mathematics Subject Classification (1991): 34B15, 34C15, 34C25, 58F10, 58F14, 58F21

ISBN 3-540-54512-3 Springer-Verlag Berlin Heidelberg New York
ISBN 0-387-54512-3 Springer-Verlag New York Berlin Heidelberg

Printed in Germany

Typesetting: Camera ready by author
Printing and binding: Druckhaus Beltz, Hemsbach/Bergstr.
2146/3140-543210 - Printed on acid-free paper

Preface

Watching the cyclic motion of the planets around the sun, the seasons on earth, our biological day and night rythm, the periodicity of life itself (to be born, live and die) are only a few examples showing that we are embedded in and surrounded by cyclic phenomena. In fact, periodic solutions of mathematical models of physical systems arise precisely because we live in a cyclic (or nearly so) world.

In studying the extensive literature on periodic and almost periodic solutions of non-linear ODE's, I think the first question comming to mind is why there should be another text on this topic. My reasons for writing are multiple. I suppose anyone who has ever studied the theory of non-linear vibrations was surprised to discover motion readily discernible by us, the periodic motion, is so difficult to investigate analytically despite highly developed mathematical tools; and in most cases there is no analytical access at all. There is a large number of mathematicans and scientists who developed analytical tools for the investigation of periodic solutions of non-linear ODE's; indeed, the field is nearly saturated. Hence, the probability is very low of making serious progress in developing further analytical methods. However, the numerical treatment of ODE's reached a very high level in the last two decades, and it seems reasonable to apply and to tailor numerical algorithms for the purpose of computation and investigation of periodic solutions. Furthermore, the bifurcation behavior of periodic "modes", due to varying parameters of a dynamical system, was found to be the most important mechanism to explain and to investigate transition into "chaos". Bifurcation of cyclic motions is caused by destabilization and therefore computation, stability and bifurcation analyses of periodic solutions are elementary steps in gaining essential information about a non-linear dynamical system.

Although periodic motions are related to many applications, most of the literature about this topic involves a very abstract mathematical framework. My intention was to connect mathematical framework and applications. The mathematical tools I employ, are based on modern applied mathematics and numerical analysis. Since this book is based primarily on work undertaken during my research activities at "Institut B für Mechanik, TU München", I was mainly concerned with technical problems.

Technical problems lead in many cases to mathematical descriptions which involve dicontinuities with respect to the state space variables of the vector field. Aside from some trivial exceptions, periodic solutions of dynamical systems with discontinuities have never really been closely investigated, either in application or in theorry. Therefore I found it useful to investigate these systems with same intensity in the form of differentiable dynamical systems.

I would like to mention that this book would never have been completed in this form if the director of the Institut B für Mechanik, Prof. Dr.–Ing. F. Pfeiffer had not provided his interest and support throughout the whole time of my stay. Particularly, his liberal attitude made it possible to investigate interdisciplinary topics like the ones discussed here. Furthermore, I am thankful for the influence of his style of research which encouraged and motivated me constantly. I also wish to thank Prof. Dr. rer. nat. R. Bulirsch for reviewing the manuscript and for his support in connecting modern applied mathematics with engineering problems. I would like to express my special thanks to Prof. Dr. rer. nat. P. Rentrop for his useful and constructive suggestions on the treatment of numerical problems. Regarding the English version I thank Prof. Dr. Dr. hc. mult. G. Leitmann for his suggestions and influence during my stay at UC Berkeley in improving the readability of this text. In addition, my thanks are due to all of my collegues who supported the work in some sense, in particular Dr.–Ing. habil. H. Bremer, Dr.–Ing. K. Karagiannis, Dipl.–Ing. K. Richter and Dipl.–Ing. A. Kunert who read the manuscript carefully. Last but not least it is a pleasure for me to mention my appreciation for the excellent work of Monika Böhnisch who typed the manuscript and of Monika Rotenburg who reviewed the English version of the manuscript.

Berkeley, March 1991 Eduard Reithmeier

Acknowledgement:
The text was translated into English during the author's stay at UC Berkeley as a Feodor Lynen Research Fellow of the Alexander v. Humboldt Foundation, Bonn, Germany.

Contents

1 Introduction

... denn das reine Chaos ist vollkommen uninteressant.
W. Heisenberg [1]

1.1 Motivation and objective

The determination and investigation of singular points and periodic solutions of non-linear vibrating systems is of theoretical interest as well as of technical importance. The theoretical interest stems from the convergence of many demanding mathematical fields such as the theory of fixed points and singularities, the theory of non-linear algebraic equations or the theory of ODE's. The technical relevance is shown in several points: limit cycles occur in numerous areas of application and – because of their structural invariant properties – they play an important role in the behavior of the non-linear vibrating system. Last but not least, the bifurcation of a limit cycle is one of the main reasons for the genesis of the irregular behavior of the system.

The dynamics of machinery is one of the main areas of application. Belts and chains for drive and control, for instance, rotate in a stable and continuous way if the number of revolutions is low. By increasing the speed the motion loses its stability and a periodic vibration will result. The same effect occurs in railway vehicle systems that have a stable straight-forward run at low speeds. If the vehicle surpasses a certain velocity, a limit cycle motion will result. This phenomen is not restricted to differentiable systems. Systems which are discontinuous with respect to state space variables also show this behavior. Examples for this case are damping elements with dry friction which are implemented between the blades of a turbine. If the number of revolutions is low, the damping elements lock. With increasing speed, however, a limit cycle vibration occurs. Another example with discontinuities in the state space are non-loaded gear wheel sets in multistage gear boxes. By increasing the speed or the amplitude of excitation or the damping, which is caused by oil, the vibration behavior is characterized by a series of bifurcations of periodic solutions.

A further interesting area of application is a well-known effect: amplifiers or microphones begin to make a whistling sound if the feedback is sufficiently high. This sound is equivalent to a limit cycle motion. Furthermore, many applications from other areas such as biology, laseroptics, quantuum-mechanics or celestial-mechanics show the occurance of limit cycle vibrations.

Limit cycle motions have been a well-known phenomen of the non-linear vibration theory for many years [POINCARÉ 1912], [HOPF 1942], [MAGNUS 1955]. The

[1] Werner Heisenberg: Schritte über Grenzen. Serie Piper, vol. 336, p.236.

systems are mainly investigated by approximation theory concerning this aspect. In most cases applications have been restricted to systems with one degree of freedom. Moreover, these systems have been investigated without discontinuities.

Hence, the targets of this book are:

1. to develop techniques or to modify existing methods to compute fixed points and periodic solutions of non-linear vibrating systems,
2. to determine criteria and resulting methods which can be used to investigate the stability- and bifurcation behavior of the numerically computed periodic solutions,
3. to obtain a connection between the periodic solutions and the global behavior of the system from the theory of normal forms of POINCARÉ and SIEGEL, and to acquire knowledge of the stability- and bifurcation behavior.

The contents of this book are divided in two main chapters: the first part is dealing with differentiable systems having no discontinuities with respect to the state space variables (chapter 2) and the second part is dealing with differentiable systems with a finite number of discontinuities (chapter 3). In both chapters the main emphasis of application lies on dynamical systems.

From the mathematical point of view fixed points are also periodic solutions, namely the trivial constant solutions with respect to time. On the other hand, periodic solutions are fixed points in a suitable POINCARÉ–section in the phase space. Therefore, it is best to begin the investigations with the classification of the fixed points or the singularities, respectively. There is an immediate connection between the singularities and the theory of normal forms of POINCARÉ and SIEGEL. Furthermore, the theory of normal forms is appropriate to create connections between the stability and the bifurcation of singularities and the diffeomorphic transformation of the non-linear system into the system linearized around the singularity. For these reasons in chapter 2.3 the essential parts of the theory of normal forms will be presented. In particular, this theory will be adapted to dynamical systems. Chapter 2.4 deals with practical aspects of classifying singularities. The classification of these singularities is demonstrated in a series of examples.

HAMILTONian systems belong to dynamical systems which have a great variety of mathematical structures. These systems are neither dissipative nor internally or externally excited. This is the reason why HAMILTONian systems play a special role and have to be treated in a separate way, which is done in chapter 2.5. As an example for HAMILTONian systems, the double pendulum is investigated. For this example in chapter 2.5.4 all periodic solutions will be computed, which are necessary to explain the irregular behavior of such systems in certain energy intervals (cf. chapter 2.8.4).

Dynamical systems which are dissipative and excited as well have less mathematical structure, but they often occur in technical problems. In such systems the limit cycle motion is dominant. Methods (such as the multiple shooting method) for computing limit cycles can be found in chapter 2.6. The efficiency of the methods will be demonstrated by a railway vehicle system with four degrees of freedom.

In many areas of application it is important to vary the system parameters to obtain periodic solutions with additional properties. One possibility is to construct a periodic solution with a minimal time of period or maximal stability. Another example is the so-called "synchronisation". By this method an excitable vibrating system will be adjusted to the frequency of a harmonic outside excitation. In chapter 2.7 a numerical algorithm, consisting of the HAMILTONian theory of optimization and the multiple shooting method, will be constructed to compute periodic solutions for this problem. The algorithm will be applied to the double pendulum and the railway vehicle system. Furthermore, a connection between the numerical computation of bifurcation points and this algorithm can be established.

Chapter 2.8 deals with the stability and bifurcation of periodic solutions. It will be shown that each solution embedded in a field of asymptotic stable solutions must converge to a limit function. This limit function is either a fixed point or a limit cycle. Two necessary conditions will be obtained for the bifurcation of periodic solutions. As above, the results are applied to the model of a railway vehicle system and double pendulum.

All results of chapter 2 can be transferred with some modifications to differentiable systems with discontinuities. Therefore, chapter 3 is divided into the same subjects as chapter 2. However, the treatment of the corresponding parts is shorter. For the application of the methods, a one-staged gear wheel set and an excitation model with dry friction will be taken into consideration.

1.2 Survey of literature

1.2.1 Existence of periodic solutions

The interest in periodic solutions of non-linear vibrating systems goes back to the beginning of this century. Already [POINCARÉ 1912] investigated periodic solutions of non-linear dynamical systems. In connection with the restricted three-body-problem he studied fixed points of area-preserving one-to-one transformations of simply connected areas on the plane. However, he could not prove his well known "last geometric problem" himself. BIRKHOFF solved this problem some years later and extended it to an arbitrary dimension of the state space. Based on this theorem, [BIRKHOFF, LEWIS 1933] have proved the existence of an infinite number of periodic solutions of a conservative system in the neighborhood of a known periodic solution, which could also be a fixed point of the "general stable type".

An interesting step was taken by [SEIFFERT 1945] twelve years later. He showed, that for each energy level of a dynamical system, described by the LAGRANGEian equations $\left(\frac{\partial T}{\partial \dot{\mathbf{q}}}\right)^{\cdot} - \frac{\partial T}{\partial \mathbf{q}} + \frac{\partial V}{\partial \mathbf{q}} = \mathbf{0}$, at least one periodic solution exists. In contrast to the theorem of BIRKHOFF, his proof was based purely on the tools of differential geometry. He was looking for closed geodesics on a RIEMANNian manifold with the metric $d\,s^2 := (H-V)d\mathbf{q}^T\mathbf{M}\,d\mathbf{q}$, where $\mathbf{M}$ is the mass matrix, H the HAMILTONian function and V the potential function of the system.

Parallel to these results, [HOPF 1942] proved a theorem which is very useful for technical problems. This theorem supplies a sufficient criterion for the existence of limit cycles in the neighborhood of a singularity of a dissipative and excited non-linear vibrating system.

In the following years mainly [SIEGEL 1954], [SIEGEL 1971] and [MOSER 1953] were occupied with a series of extensions of the theorem of BIRKHOFF and the so-called "resonant case".

Later in the 1960's and '70's many authors such as [HARRIS 1966], [BERGER 1971], [GORDON 1971], [RABINOWITZ 1978a], [EKELAND 1979], [AMANN, ZEHNDER 1980] etc., dealt with periodic solutions of HAMILTONian systems. For special classes of these systems they obtained theorems about the existence of periodic solutions. The main idea of these proofs was to transform the problem of finding periodic solution into a minimax problem of the calculus of variation. By doing this, it turns out that for each periodic solution a critical point of the corresponding variational problem exists.

Other approaches – coming from the differential topology – were made by [MARKUS 1960], [FULLER 1967] and [WEINSTEIN 1973]. Analogously to the topological index in the theory of fixed points, FULLER defined an index for periodic solutions of autonomous systems and applied it to prove the existence of periodic solutions of HAMILTONian systems. In 1973 WEINSTEIN showed that on each energy surface of a HAMILTONian system $\dot{\mathbf{z}} = \mathbf{J}\cdot DH(\mathbf{z})$ ($\mathbf{J}$ symplectic) at least f periodic solutions exist which are seperated. The energy surface lies in a certain neighborhood of a singularity $\mathbf{z}_0 \in \mathbb{R}^{2f}$ and furthermore the HESSE–matrix $D^2H(\mathbf{z}_0)$ must be positive definite. [MOSER 1976] extended this result to arbitrary systems $\dot{\mathbf{z}} = \mathbf{f}(\mathbf{z})$ which have a first integral.

A survey of the existence-statements of periodic solutions of non-linear dynamical systems may be found in [YOSHIZAWA 1975], [RABINOWITZ 1982] and [DUISTERMAAT 1984] for instance.

1.2.2 Numerical computation of periodic solutions

An analytical expression of a periodic solution is mostly restricted to trivial cases. Therefore, up to the 60's the investigations were concentrated on the existence or uniqueness, or on the approximation methods (cf. [MAGNUS 1955]). Since POINCARÉ published his results periodic solutions have mainly been used in the field of pure mathematics. Also the early numerical treatment by the analogue-computer for the DUFFING–oscillator, the VAN DER POL–oscillator or other examples with one degree of freedom, did not change the situation.

A new era has started in the 80's: numerical algorithms were employed to compute and investigate periodic solutions. [SEYDEL 1983] as well as [HOLODNIOK, KUBIČEK 1984] used the multiple shooting method developed by [BULIRSCH, STOER, DEUFLHARD 1977]. The idea is based on finding the periodic solution by solving a boundary value problem with an unknown time of period. Because of the non-uniqueness of the periodic solutions in the autonomous case, it is problematic to formulate the boundary-value problem. Therefore – based on the multiple shooting method – [DEUFLHARD 1984] generated a modified GAUSS–NEWTON–technique to get rid of this problem.

Another interesting proposal – based on the index theory of POINCARÈ–BENDIXON – came from [HSU 1980a,b]. The idea is to generate a POINCARÈ–mapping $P : \Sigma \to \Sigma$ ($\Sigma \subset TM$, $codim\, \Sigma = 1$) in the state space TM. Then the periodic solution is equivalent to a fixed point of P. Now, if a JORDAN–curve $\Gamma : S^{n-1} \to \Sigma$ is continuously deformed, the index of $P - id_{\Sigma}$ changes from $+1$ to -1 or vice versa if the curve Γ passes a fixed point of P.

1.2.3 Bifurcation and stability of periodic solutions

The investigation of stability and bifurcation requires the periodic solutions either in analytical or numerical form. The first investigations in this area were of pure analytical nature. Especially the criteria of [HOPF 1955] should be mentioned. They suppply a condition for the bifurcation from a fixed point to a periodic solution.

Among others, analytical investigations were made mainly by [MEYER 1970], [MEYER 1971], [MIL'SHTEIN 1977], [BOTTKOL 1977] and [MÜLLER 1981]. MEYER investigated the bifurcation of periodic points (fixed points) of a vector field on a two-dimensional manifold which depends on one parameter. Based on these results, he classified the periodic points. Furthermore, he analysed the stability properties of each of these classes by the invariant curve-theorem of MOSER. By doing this he used stability criteria based on LIAPUNOV–functions. Also based on LIAPUNOV–functions, however in connection with optimization techniques, MIL'SHTEIN obtains criteria for the asymptotic stability of periodic solutions. MÜLLER modifies these stability theorems to apply them to limit cycles which are computed approximatively. However, he restricted his investigations to point-symmetric vector fields. BOTTKOL deals with vector fields, for which a parameter-dependent submanifold exists in the state space. His problem formulation concerns structural stability. That means, he investigates a vector field in the "neighborhood" and looks for a neighboring parameter-dependent submanifold of periodic solutions.

The investigation of bifurcation and stability of periodic solutions, which are numerically computed, can especially be found in [SEYDEL 1988].

1.2.4 Periodic solutions of dynamical systems with discontinuities

Already [SENATOR 1969] investigated stability and bifurcation aspects of periodic solutions with a one degree of freedom system subjected to impacts. Similar examples can be found in [HSU 1977] and [HARTOG 1931]. Since there is an analytical solution of these examples, the investigation is mainly focused on the special situation.

In the last few years, the number of authors dealing with numerical analysis of nonlinear systems with discontinuities has risen. Here particularly [HOLMES 1982], [SHAW 1985], [HEIMANN, BAJAJ, SHERMAN 1988], [PFEIFFER 1988a,b], [KARAGIANNIS 1989] and [MEIJAARD, DE PATER 1989] have to be mentioned. In these works the mathematical modelling, numerical computation and simulation are considered. Some authors such as HEIMANN et al. and SHAW also discuss the stability and bifurcation behavior of periodic solutions with discontinuities.

2 Differentiable dynamical systems

"The study of these cyclic or periodic vibrations is the study of vibration and it is one of the most important aspects of dynamics."

R.F. Steidel [2]

2.1 Preliminary remarks

The frame for the investigation – that is the numerical computation, stability and bifurcation – of singularities and periodic solutions are differentiable vector fields

$$\mathbf{f}^* : TM \times U \times P \times I \rightarrow TM \tag{2.1}$$

whose trajectory $\Phi_\xi : [0, \infty[\rightarrow TM$ with the initial point $\boldsymbol{\xi} \in TM$ is given for each time t by the unique initial value problem

$$\begin{aligned} \dot{\mathbf{x}}(t) &= \mathbf{f}^*(\mathbf{x}(t), \mathbf{u}(t), \mathbf{p}, t) \; , \\ \mathbf{x}(0) &:= \boldsymbol{\xi} \; . \end{aligned} \tag{2.2}$$

In this formulation

M : is the configuration space (differentiable manifold, locally isomorphic to $\mathbb{R}^f$), $\dim M = f \in \mathbb{N}$, f is the number of degrees of freedom of the system,

TM : $= M \times \mathbb{R}^f$ is the state space,

$U \subset \mathbb{R}^m$: is the range of values of the control $\mathbf{u}$, which is mostly obtained by an optimization strategy or a control design,

$P \subset \mathbb{R}^k$: is the space of parameters, which can be varied in the system,

$I \subset \mathbb{R}$: is the range of time of excitation, which is mostly given by an explicit time function.

---

[2] From [STEIDEL 1989], page 40.

The vector field $\mathbf{f}^*$ is assumed to be sufficiently differentiable. The partial differentiability is related to the variables $\mathbf{x}, \mathbf{u}, \mathbf{p}$ and t.

∇

Example 2.1: Railway vehicle system.

Fig. 2.1 shows a simple dynamical model of a wheel set of a railway-vehicle system. The (rigid) wheel set (double cone with cone-angle δ) is elastically mounted in the undercarriage. The number of degrees of freedom is 6.

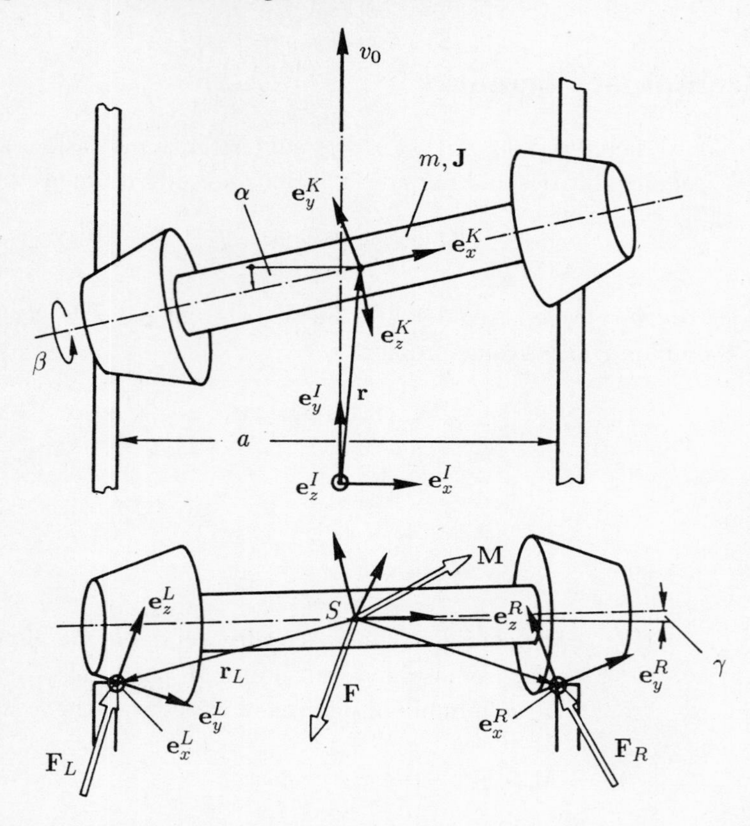

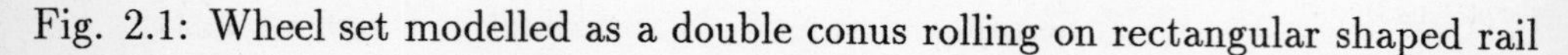

Fig. 2.1: Wheel set modelled as a double conus rolling on rectangular shaped rail

To investigate the dynamical behavior of the wheel set we assume that the undercarriage runs straight forward at a constant velocity $\mathbf{v}$. The feedback from the wheel set to the undercarriage is neglected because of the higher mass of the undercarriage.

The position of the wheel set (with respect to an inertial frame $I := \left(\mathbf{e}_x^I, \mathbf{e}_y^I, \mathbf{e}_z^I\right)$, which is moved at a constant velocity v_0 of the wheel suspension) is described by the coordinates x, y and z as well as by the cardanian angles α, β and γ. Hence, the configuration space is given by

$$M = \mathbb{R}^3 \times SO(3) \ .$$

The state space or phase space

$$TM = M \times \mathbb{R}^6$$

contains all positons $(\mathbf{r}, \mathbf{T}) \in M$ ($\mathbf{r} := (x, y, z)^T \in \mathbb{R}^3$, $\mathbf{T} := \mathbf{T}_\gamma \circ \mathbf{T}_\beta \circ \mathbf{T}_\alpha \in SO(3)$; $\mathbf{T}_\alpha, \mathbf{T}_\beta, \mathbf{T}_\gamma$ are rotations around the angles α, β, γ) and all velocities $(\mathbf{v}, \boldsymbol{\omega}) \in \mathbb{R}^6$. $\mathbf{v} \in \mathbb{R}^3$ is the velocity of the center S of the mass of the wheel set. $\boldsymbol{\omega} \in \mathbb{R}^3$ is its (absolute) angular velocity.

With respect to the inertial frame I, the velocity $\mathbf{v}$ has the representation ${}_I\mathbf{v} := (\dot{x}\ ,\ v_0 + \dot{y}\ ,\ \dot{z})^T$. The angular velocity $\boldsymbol{\omega}$ or $\tilde{\boldsymbol{\omega}}$ respectively has the representation ${}_K\tilde{\boldsymbol{\omega}} := \mathbf{T}^T\dot{\mathbf{T}}$ with respect to the body-fixed frame $K := \left(\mathbf{e}_x^K, \mathbf{e}_y^K, \mathbf{e}_z^K\right)$, if α, β and γ define the angles of rotation between the frames K and I. $U \subset \mathbb{R}^m$ is the control space (forces/moments used to control the system), if the wheel set is controlled.

The following parameters might be used:

- the reference-velocity v_0,
- the mass m and the matrix of the moments of inertia $\mathbf{J}$ of the wheel set,
- the radius of the wheels in a stationary drive ($\mathbf{r}(t) = 0\ ,\quad \mathbf{T}(t) = \mathbf{E}_3$),
- the cone-angle δ,
- the gauge of the rails,

  ⋮

  etc.

$\Rightarrow \quad \mathbf{p} := (v_0\ ,\ a\ ,\ r_0\ ,\ \ldots) \in P$.

The range of the parameter values is mostly restricted because of technical reasons: $P \subset \mathbb{R}^k$. k is the number of the parameters p_i.

The state vector $\mathbf{x}$ is given by

$$\mathbf{x} := (\,{}_I\mathbf{r}\ ,\ \mathbf{T}\ ,\ {}_I\mathbf{v}\ ,\ {}_K\tilde{\boldsymbol{\omega}})\ .$$

The equations of motion, expressed in state space variables, are

$$\begin{aligned}
{}_I\dot{\mathbf{r}} &= {}_I\mathbf{v}\ ,\\
\dot{\mathbf{T}} &= \mathbf{T} \cdot {}_K\tilde{\boldsymbol{\omega}}\ ,\\
{}_I\dot{\mathbf{v}} &= \tfrac{1}{m}\left({}_I\mathbf{F} + {}_I\mathbf{F}_L + {}_I\mathbf{F}_R\right)\ ,\\
{}_K\dot{\boldsymbol{\omega}} &= {}_K\mathbf{J}_S^{-1}\left({}_K\mathbf{M}_S - {}_K\tilde{\boldsymbol{\omega}}\, {}_K\mathbf{J}_S\, {}_K\boldsymbol{\omega} + {}_K\mathbf{r}_L \times {}_K\mathbf{F}_L + {}_K\mathbf{r}_R \times {}_K\mathbf{F}_R\right)
\end{aligned}$$

or

$$\dot{\mathbf{x}} = \mathbf{f}^*(\mathbf{x}\ ,\ \mathbf{u}\ ,\ \mathbf{p}\ ,\ t)\ ,$$

respectively.

And, according to Fig. 2.1:

- ${}_I\mathbf{F}_L$ (${}_I\mathbf{F}_R$) denote the contact forces between the left (right) cone-surface and the rails represented in the frame I,
- ${}_I\mathbf{r}_L$ (${}_I\mathbf{r}_R$) denote the position vectors from the center S of the mass of the wheel set to the left (right) point of contact,
- (${}_I\mathbf{F}$, ${}_K\mathbf{M}$) denote the equivalent pair of forces/torques with respect to S acting between wheel suspension and wheel set,
- ${}_K\mathbf{J}_S$ denotes the representation of the tensor of the moments of inertia of the wheel set in the frame K with respect to S.

△

If a (non-linear) control scheme or if geometrical constraints are given, additional non-linear algebraic constraints

$$\mathbf{g}(\mathbf{x}\,,\,\mathbf{u}\,,\,\mathbf{p}\,,t) = \mathbf{0} \tag{2.3}$$

have to be satisfied. If the system is completely controlable, it is appropriate to assume that the algebraic constraints (2.3) are eliminated because in this case for each $(\mathbf{x}, \mathbf{p}, t) \in TM \times P \times \mathbb{R}$ there is a locally unique $\mathbf{u} \in U$, say

$$\mathbf{u} = \boldsymbol{\psi}(\mathbf{x}, \mathbf{p}, t) \quad . \tag{2.4}$$

Equation (2.4) together with equation (2.2) supplies the vector field

$$\mathbf{f}(\mathbf{x}, \mathbf{p}, t) := \mathbf{f}^*(\mathbf{x}\,,\,\boldsymbol{\psi}(\mathbf{x}, \mathbf{p}, t)\,,\,\mathbf{p}\,,\,t) \quad , \tag{2.5}$$

which is unique in a neighborhood of each point $(\mathbf{x}, \mathbf{p}, t)$.

Numerical algorithms for differential-algebraic-equations (DAE) with applications can be found in [REITHMEIER 1984], [GRIEPENTROG, MÄRZ 1986], [PETZOLD, LÖTSTEDT 1986], [FÜHRER, C. 1988]. In case, the bodies of the rigid body system are connected and suspended against its environment, we have an algebraic constraint which can be expressed by

$$\mathbf{h}(\mathbf{x}, \mathbf{p}, t) = \mathbf{0} \quad . \tag{2.6}$$

And the equation of motion (2.5) according to the condition (2.6) may be formulated by a DAE:

$$\begin{aligned} \dot{\mathbf{x}} &= \mathbf{f}(\mathbf{x}, \mathbf{p}, t) + [D\mathbf{h}(\mathbf{x}, \mathbf{p}, t)]^T \boldsymbol{\lambda} \ , \\ \dot{t} &= 1 \ , \\ 0 &= \mathbf{h}(\mathbf{x}, \mathbf{p}, t) \end{aligned} \tag{2.7}$$

with

$$D\mathbf{h}(\mathbf{x},\mathbf{p},t) := [D_x\mathbf{h}(\mathbf{x},\mathbf{p},t) \ , \ D_t\mathbf{h}(\mathbf{x},\mathbf{p},t)] \ . \tag{2.8}$$

The vector $\boldsymbol{\lambda}$ denotes the LAGRANGE multipliers which generally depend on the time t.

Equation (2.7) can be interpreted in the following way: by employing the scalar product between an arbitrary point $\dot{\bar{\mathbf{x}}}$ of the tangent space TM and the dynamical system (2.5), the equation

$$\dot{\bar{\mathbf{x}}}^T(\dot{\mathbf{x}} - \mathbf{f}(\mathbf{x},\mathbf{p})) = 0 \quad [3] \tag{2.9}$$

must be valid for each solution of the ODE $\dot{\mathbf{x}} = \mathbf{f}(\mathbf{x},\mathbf{p})$. If $\dot{\bar{\mathbf{x}}}$ is a point which is simultanously a solution of the homogenous system of equations

$$D\mathbf{h}(\mathbf{x},\mathbf{p})\dot{\bar{\mathbf{x}}} = \mathbf{0} \ , \tag{2.10}$$

for instance $\dot{\bar{\mathbf{x}}} = \mathbf{X}(\mathbf{x},\mathbf{p}) \cdot \dot{\bar{\boldsymbol{\eta}}}$ with $\dot{\bar{\boldsymbol{\eta}}} \in \mathbb{R}^r$, $r = \dim(KerD\mathbf{h}(\mathbf{x},\mathbf{p}))$, then equation (2.9) supplies

$$\mathbf{X}(\mathbf{x},\mathbf{p})^T \cdot (\dot{\mathbf{x}} - f(\mathbf{x},\mathbf{p})) = 0 \ . \tag{2.11}$$

Equation (2.11) is valid for each solution of $\dot{\mathbf{x}} = \mathbf{f}(\mathbf{x},\mathbf{p})$ which additionally satisfies the constraints (2.6).

Because of

$$D\mathbf{h}(\mathbf{x},\mathbf{p}) \cdot \mathbf{X}(\mathbf{x},\mathbf{p}) = \mathbf{0} \tag{2.12}$$

it turns out equations (2.7) and (2.11) are identical. The LAGRANGE–multipliers are equivalent to the forces to keep the motion on the surface $\mathbf{h}^{-1}(0)$.

A transformation into surface oriented coordinates

$$\mathbf{h}^{-1}(0) \ \rightarrow \ M \ , \quad \boldsymbol{\eta} \ \mapsto \ \mathbf{x} \tag{2.13}$$

with

$$\dot{\mathbf{x}} = \mathbf{X}(\mathbf{x},\mathbf{p})\dot{\boldsymbol{\eta}} \tag{2.14}$$

results in a reduced ODE–system in standard-form

$$\begin{aligned} \dot{\boldsymbol{\eta}} &= \mathbf{X}^+(\mathbf{x}(\boldsymbol{\eta}) \ , \mathbf{p}) \ \mathbf{f}(\mathbf{x}(\boldsymbol{\eta}) \ , \mathbf{p}) \ , \\ &=: \hat{\mathbf{f}}(\boldsymbol{\eta},\mathbf{p}) \end{aligned} \tag{2.15}$$

which is equivalent to equation (2.5) ($\mathbf{X}^+ :=$ pseudoinverse matrix of $\mathbf{X}$).

For rigid multi-body systems with n bodies, these equations can be expressed in a more precisely form. Namely, if

[3] Without restriction, the explicit time dependency need not be taken into account.

- $\mathbf{r} := \left({}_I\mathbf{r}^{(1)}, \ldots, {}_I\mathbf{r}^{(n)} \right) \in \mathrm{IR}^{3n}$ is the vector of the position vectors $\mathbf{r}^{(i)}$ from the origin of an inertial frame I to the center of mass C_i of the body (i); $(i = 1, \ldots, n)$, represented in the frame I,
- $\mathbf{T} := diag\left\{\mathbf{T}^{(1)}, \ldots, \mathbf{T}^{(n)}\right\}$ is the blockdiagonal rotation matrix ($\mathbf{T}^{(i)} \in SO(3)$ transforms the representation of an arbitrary vector given in frame K_i into an inertial frame),
- $\mathbf{v} := \left({}_I\mathbf{v}^{(1)}, \ldots, {}_I\mathbf{v}^{(n)} \right) \in \mathrm{IR}^{3n}$ is the vector of the (absolute) velocity $\mathbf{v}^{(i)}$ of center C_i of body (i) (represented in frame I) ; $(i = 1, \ldots, n)$,
- $\boldsymbol{\omega} := \left({}_{K_1}\boldsymbol{\omega}^{(1)}, \ldots, {}_{K_n}\boldsymbol{\omega}^{(n)} \right) \in \mathrm{IR}^{3n}$ is the vector of the (absolute) angular–velocities $\boldsymbol{\omega}^{(i)}$ of body (i) (represented in frame K_i) ; $(i = 1, \ldots, n)$,
- $\tilde{\boldsymbol{\omega}} := diag\left\{\tilde{\boldsymbol{\omega}}^{(1)}, \ldots, \tilde{\boldsymbol{\omega}}^{(n)}\right\} \in \mathrm{IR}^{3n,3n}$,
- $\mathbf{y} := (\mathbf{r}, \mathbf{T})$; $\mathbf{z} := (\mathbf{v}, \boldsymbol{\omega})$; $\mathbf{x} := (\mathbf{y}, \mathbf{z})$,
- $\mathbf{m} := diag\left\{m^{(1)}\mathbf{E}_3, \ldots, m^{(n)}\mathbf{E}_3\right\} \in \mathrm{IR}^{3n,3n}$ is the blockdiagonal mass matrix ($m^{(i)} :=$ mass of body (i)),
- $\mathbf{J} := diag\left\{ {}_{K_1}\mathbf{J}^{(1)}, \ldots, {}_{K_n}\mathbf{J}^{(n)} \right\} \in \mathrm{IR}^{3n,3n}$ is the blockdiagonal matrix of moments of inertia (${}_{K_i}\mathbf{J}^{(i)} :=$ tensor of moments of inertia of body (i) represented in frame K_i),
- $\mathbf{M} := diag\,\{\mathbf{m}, \mathbf{J}\}$,
- $\mathbf{F} := (\mathbf{F}_F, \mathbf{F}_M) := \left({}_I\mathbf{F}^{(1)}, \ldots, {}_I\mathbf{F}^{(n)} \;,\; {}_{K_1}\mathbf{M}^{(1)}, \ldots, {}_{K_n}\mathbf{M}^{(n)} \right)$ is the vector of the equivalent pairs $\left(\mathbf{F}^{(i)}, \mathbf{M}^{(i)}\right)$ of forces and torques (represented in frame I or K_i respectively) acting on body (i),

then for the multibody system the equation of motion is given by

$$\begin{aligned} \dot{\mathbf{r}} &= \mathbf{v} \;, \\ \dot{\mathbf{T}} &= \mathbf{T}\tilde{\boldsymbol{\omega}} \;, \\ \dot{\mathbf{v}} &= \mathbf{m}^{-1}\mathbf{F}_F \;, \\ \dot{\boldsymbol{\omega}} &= \mathbf{J}^{-1}\left(\mathbf{F}_M - \tilde{\boldsymbol{\omega}}\,\mathbf{J}\,\boldsymbol{\omega}\right) \end{aligned} \tag{2.16}$$

or

$$\begin{aligned} \dot{\mathbf{y}} &= \mathbf{z} \;, \\ \dot{\mathbf{z}} &= \mathbf{M}^{-1}\left(\mathbf{F} - \begin{bmatrix} \mathbf{0} \\ \tilde{\boldsymbol{\omega}}\,\mathbf{J}\,\boldsymbol{\omega} \end{bmatrix}\right) \;, \end{aligned} \tag{2.17}$$

respectively.

According to equation (2.9) the scalar product results in

$$\overline{\mathbf{z}}^T(\dot{\mathbf{y}} - \dot{\mathbf{z}}) + \dot{\overline{\mathbf{z}}}^T\left(\mathbf{M}^{-1}\left(\mathbf{F} - \begin{bmatrix} \mathbf{0} \\ \tilde{\omega}\,\mathbf{J}\,\omega \end{bmatrix}\right)\right) = 0 \ . \tag{2.18}$$

In contrast to the principle of JOURDAIN, additionally "admissable states of acceleration" will be taken into account in equation (2.9). Hence, the constraints may include terms of second time derivative or may be non-holonomous.

∇

Example 2.2: Wheel set on a rigid rail

If the wheel set in example 2.1, Fig. 2.1, moves along on the edge of a rigid rail with permanent contact, then the left (right) inner side of the rail is tangent to the left (right) cone. The condition of contact is:

$$\mathbf{T}^T\begin{bmatrix} a - x \\ \lambda - y \\ -(r_0 + z) \end{bmatrix} = \begin{bmatrix} a + \frac{r_0 - \sqrt{y_R^2 + z_R^2}}{tg\delta} \\ y_R \\ z_R \end{bmatrix} \ . \tag{A}$$

${}_K\mathbf{r}_R = (x_R\ ,\ y_R\ ,\ z_R)^T$ is the position vector from the center of mass S to the right point of contact on the rail with respect to frame K. The condition of the rail which is tangent to the cone has the form:

$$\begin{bmatrix} -\frac{\mu y_R + \nu z_R}{tg\delta\sqrt{y_R^2 + z_R^2}} \\ \mu \\ \nu \end{bmatrix} = \mathbf{T}^T\begin{bmatrix} 0 \\ 1 \\ 0 \end{bmatrix} \ . \tag{B}$$

λ, μ and ν are unknown parameters, which have to be determined from the equations (A) and (B). Additionally, employing the same equations the coordinates y_R and z_R must be computed. There is one redundant equation establishing a geometrical connection between the coordinates $(x, y, z, \alpha, \beta, \gamma)$. An analogous equation results considering the left rail. Therefore, the wheel set has only four degrees of freedom.

For small angles α and γ $\left(\mathbf{T}^T \doteq \begin{bmatrix} 1 & \alpha + \gamma\sin\beta & -\gamma\cos\beta \\ -\alpha\cos\beta & \cos\beta & \sin\beta \\ \gamma + \alpha\sin\beta & -\sin\beta & \cos\beta \end{bmatrix}\right)$ the equations (A) and (B) can be explicitly solved to obtain the parameters $(x_R, y_R, \lambda, \mu, \nu)$. According to equation (2.6) the following algebraic constraint results for the right rail:

$$h_1(\mathbf{r}, \mathbf{T}) \doteq (\gamma - \alpha\sin\beta)\left[(a - x) - (r_0 + z)\,tg\delta\right] + x\,tg\,\delta - z \ .$$

The constraint equation for the left rail can be determined in the same way. The result is:

$$h_2(\mathbf{r}, \mathbf{T}) \doteq (\gamma + \alpha \sin \beta) \left[(r_0 + z)\, tg\delta - (a + x) \right] + x\, tg\, \delta + z \ .$$

Both equations can be written in the standard vector form:

$$\mathbf{h}(\mathbf{r}, \mathbf{T}) = \begin{bmatrix} h_1(\mathbf{r}, \mathbf{T}) \\ h_2(\mathbf{r}, \mathbf{T}) \end{bmatrix} = \mathbf{0} \ .$$

For any shape of the wheels and rails different but equivalent algebraic constraints would result. This property is not essential and will be neglected.

△

The following considerations are based on the standard vector field according to equation (2.5). That means it is assumed that – without restrictions – all algebraic constraints (such as a controller or geometrical constraints) are eliminated.

Since non-autonomous ODE's may be transformed into an autonomous system by a trivial modification, it is sufficient to consider just an autonomous ODE. Results or properties which in particular are related to non-autonomous systems will be treated explicitly in other chapters. Hence, the vector field $\mathbf{f}$ of the dynamical system depends only on the state $\mathbf{x}$ and the parameter vector $\mathbf{p}$.

2.2 Some basic structures and properties of the equation of motion of dynamical systems

In most cases, the mathematical description of a dynamical model shows specific characteristics. The most important properties will be mentioned in this chapter.

2.2.1 Periodicity of configuration space coordinates

In many dynamical systems the coordinates $q_1, \ldots, q_f$ ($\mathbf{x} := q_1, \ldots, q_f,\ \dot{q}_1, \ldots, \dot{q}_f$) describing the configuration space denote angles and because of that reason they cause a 2π–periodicity in the equations of motion

$$\mathbf{f}(\ldots, x_i + 2\pi, \ldots) = \mathbf{f}(\ldots, x_i, \ldots) \ . \tag{2.19}$$

If the 2π–periodicity holds for the coordinates $x_{i_1}, \ldots, x_{i_S}$ ($s \leq f$) (without any restriction $(i_1, \ldots, i_s) = (1, \ldots, s)$), then an unbounded configuration space is denoted by

$$M \tilde{=} T^s \times \mathbb{R}^{f-s} \tag{2.20}$$

with

$$T^s := \underbrace{S^1 \times \ldots \times S^1}_{s\text{–mal}} \ . \tag{2.21}$$

2.2.2 Analyticity of the dynamical system

Mostly a C^∞–function $\mathbf{f}$ with respect to the coordinates of the state space describs the right hand side of the equation of motion of the dynamical system. Therefore and because of equation (2.19), the vector field $\mathbf{f}$ may be expressed by a combined TAYLOR– and FOURIER–expansion:

$$\mathbf{f}(\mathbf{x}, \mathbf{p}) = \sum_{\mu,\nu,\kappa} \mathbf{C}_{\mu\nu\kappa}(\mathbf{p})\, \mathbf{v}^{\mu}\, e^{i<\nu,\mathbf{w}>}\, \mathbf{e}_\kappa \ . \tag{2.22}$$

According to equation (2.20) the state space vector $\mathbf{x}$ consists of the parts

$$\begin{aligned} \mathbf{v} &:= (x_{s+1}, \ldots, x_n) \ , \\ \mathbf{w} &:= (x_1, \ldots, x_s) \ . \end{aligned} \tag{2.23}$$

The vectors

$$\begin{aligned} \boldsymbol{\mu} &:= (\mu_{s+1}, \ldots, \mu_n) \in \mathbb{Z}^{n-s} \ , \\ \boldsymbol{\nu} &:= (\nu_1, \ldots, \nu_s) \in \mathbb{Z}^s \end{aligned} \tag{2.24}$$

denote multi indices. Furthermore the following definitions are used for the sake of clarity.

$$
\begin{aligned}
\mathbf{v}^{\mu} &:= v_1^{\mu_1} \cdot \ldots \cdot v_s^{\mu_s} \ , \\
< \boldsymbol{\nu}, \mathbf{w} > &:= \sum_{k=1}^{s} \nu_k w_k \ , \\
\mathbf{C}_{\mu\nu\kappa}(\mathbf{p}) &:= \tfrac{1}{2\pi} \int_Q \mathbf{f}(\mathbf{x}, \mathbf{p}) \, e^{-i<\nu,\mathbf{w}>} d\mathbf{w} \ , \\
Q &:= [0, 2\pi]^s \ , \\
\mathbf{e}_\kappa &:= (0, \ldots, \underset{\underset{\kappa}{\uparrow}}{1}, \ldots, 0)^T \ .
\end{aligned}
\tag{2.25}
$$

The analyticity of $\mathbf{f}$ is a necessary condition in the theory of normal forms of vector fields. Additionally, in many cases the analyticity of $\mathbf{f}$ provides the computation or estimation respectively of the FOURIER–coefficients of the periodic solutions of equation (2.2) [MICKENS 1988].

2.2.3 Representation in state space coordinates

In most cases of application, the equation of motion is given by equation (2.26).

$$
\begin{aligned}
\dot{\mathbf{x}} &= \begin{bmatrix} \dot{\mathbf{q}} \\ \mathbf{a}^*(\mathbf{q}, \dot{\mathbf{q}}, \mathbf{p}) \end{bmatrix} + \begin{bmatrix} \mathbf{0} \\ \mathbf{B}(\mathbf{q}, \dot{\mathbf{q}}, \mathbf{p}) \end{bmatrix} \mathbf{u} \\
&= \mathbf{f}^*(\mathbf{x}, \mathbf{u}, \mathbf{p}) \ .
\end{aligned}
\tag{2.26}
$$

In this representation, $\mathbf{B} : TM \times P \rightarrow \mathbb{R}^{f,m}$ is the control matrix, and $\mathbf{a}^* : TM \times P \rightarrow \mathbb{R}^f$ is the vector-function of the generalized acceleration. $\mathbf{f}^*$ is linear with respect to the coordinates $(x_{f+1}, \ldots, x_{2f})$ as well as to the control vector $\mathbf{u}$, which represents the forces and the actuator torques of the dynamical system. The linearity of $f_1^*, \ldots, f_f^*$ with respect to $x_{f+1}, \ldots, x_{2f}$ will not be influenced, if – according to equation (2.3) – the control vector $\mathbf{u}$ is replaced by a state space feedback. Namely, the elimination of $\mathbf{u}$ by means of equation (2.4) and (2.26) results in:

$$
\dot{\mathbf{x}} = \begin{bmatrix} x_{f+1} \\ \vdots \\ x_{2f} \\ ----- \\ \mathbf{a}(\mathbf{x}, \mathbf{p}) \end{bmatrix} =: \mathbf{f}(\mathbf{x}, \mathbf{p}) \tag{2.27}
$$

with

$$\mathbf{a}(\mathbf{x},\mathbf{p}) := \mathbf{a}^*(\mathbf{x},\mathbf{p}) + \mathbf{B}(\mathbf{x},\mathbf{p})\,\boldsymbol{\psi}(\mathbf{x},\mathbf{p}) \ . \tag{2.28}$$

2.2.4 Singular points

Besides the periodic solutions of the dynamical system $\dot{\mathbf{x}} = \mathbf{f}(\mathbf{x},\mathbf{p})$, which are investigated in this book, the "singular points" (fixed points, points of equilibrium, stationary points) of the vector field $\mathbf{f}$ are one of the most important features to investigate the dynamical behavior of the system. The singular points are defined by the zeros of the field $\mathbf{f}$:

$$(\mathbf{x}_0,\mathbf{p}) \text{ singular point } :\Longleftrightarrow \mathbf{f}(\mathbf{x}_0,\mathbf{p}) = 0 \ . \tag{2.29}$$

Bifurcations excluded, the singular points should be isolated, if $\mathbf{p} \in P$ is fixed. That means for each singular point of $\mathbf{f}$ an – open – neighborhood of $\mathbf{x}_0$ exists, which does not contain any other singular points if $\mathbf{p} \in P$ is fixed.

Inside that neighborhood, $\mathbf{x}_0 \in TM$ may be expressed by a function $\boldsymbol{\sigma} : P \to TM$, $\mathbf{p} \mapsto \mathbf{x}_0$ which is continuous with respect to $\mathbf{p}$, if

$$rk\,[D\mathbf{f}\,(\boldsymbol{\sigma}(\mathbf{p}),\mathbf{p})] = 2f \tag{2.30}$$

is true (implicit function theorem).

By means of a linear map $\boldsymbol{\tau} : \mathbf{x} \mapsto \mathbf{x} + \mathbf{x}_0$, the vector field $\mathbf{f}$ will be transformed into a – diffeomorph – field

$$\tilde{\mathbf{f}} := \mathbf{f} \circ \boldsymbol{\tau} \tag{2.31}$$

with the singular point $(\mathbf{0},\mathbf{p})$. Hence, without any restriction we may assume that the singular point is given by $(\mathbf{x}_0,\mathbf{p}) = (\mathbf{0},\mathbf{p})$.

The existence and characterization of singularities or fixed points play an important role for the existence and the behavior of the periodic solutions of the system, which are necessary to investigate the global behavior of the system (chapter 2.3).

Because of that reason, any investigation of the dynamical behavior of the system the classification of the singular points of the vector field $\mathbf{f}$ is needed.

2.3 Normal forms of non–resonant vector fields

... In die Tiefe mußt du steigen,
Soll sich dir das Wesen zeigen.
Nur Beharrung führt zum Ziel,
Nur die Fülle führt zur Klarheit
und im Abgrund wohnt die Wahrheit.

F. Schiller [4]

In this chapter we will assume that the dynamical system $\mathbf{f}$ is analytic and there are singular points or periodic solutions. Provided that it is possible – in many cases and especially in dissipative and excited systems – to transform the non-linear system into a linear system. That linear system is the linearized part of the non-linear system about the singular points or the periodic solution. The transformation employed is usually represented by a series expansion which – in general – cannot be expressed by elementary functions.

Therefore the investigation is more of qualitative nature. Nevertheless the transformation may be useful from a practical point of view, because inside the area of convergence of the series the behavior of the non-linear system is diffeomorph to its linear part. The idea of that transformation can be put down to [POINCARÉ 1893] and [DULAC 1908] and was improved – with respect to the behavior of convergence – by [SIEGEL 1954]. In [ARNOLD 1983] a detailed investigation about this subject is made. Unfortunately the survey does not take into account the special structure of dynamical systems (equation (2.27)).

In order to transform a complete controllable dynamical system into its normal form, we assume – without any restriction – that the algebraic constraints are eliminated, because in that case for each $(\mathbf{x}, \mathbf{p}) \in TM \times P$ the state feedback supplies a local unique $\mathbf{u} \in U$.

In the neighborhood of a singular point $(\mathbf{x}_0, \mathbf{p}) = (\mathbf{0}, \mathbf{p})$ the vector field $\mathbf{f}$ has the following form:

$$\begin{aligned} \mathbf{f}(\mathbf{x}, \mathbf{p}) &= D\mathbf{f}(\mathbf{0}, \mathbf{p})\mathbf{x} + \dots , \\ &=: \mathbf{A}(\mathbf{p})\mathbf{x} + \dots . \end{aligned} \tag{2.32}$$

If

$$\mathbf{a}(\mathbf{x}, \mathbf{p}) = D\mathbf{a}(\mathbf{0}, \mathbf{p})\mathbf{x} + \sum_{i \geq 2} \mathbf{a}_i(\mathbf{x}, \mathbf{p}) \tag{2.33}$$

is the TAYLOR–expansion of $\mathbf{a}$ – according to equation (2.28) – at the singularity $(\mathbf{0}, \mathbf{p})$, then

[4] Saying of Confuzius: from the complete works of Schiller vol. I p. 369, Gotta'scher Verlag Tübingen, 1853

$$\mathbf{f}(\mathbf{x},\mathbf{p}) = \underbrace{\begin{bmatrix} \mathbf{0} & \mathbf{E} \\ D\mathbf{a}(\mathbf{0},\mathbf{p}) \end{bmatrix}}_{=:\mathbf{A}(\mathbf{p})} \mathbf{x} + \sum_{i\geq 2} \begin{bmatrix} \mathbf{0} \\ \mathbf{a}_i(\mathbf{x},\mathbf{p}) \end{bmatrix} \tag{2.34}$$

is true for dynamical systems. $\mathbf{a}_i$ contains all monomials $\mathbf{x}^\mu$ of degree i ($\mid \boldsymbol{\mu} \mid = i$). The basic idea of transforming the system into its normal form is that all polynomials $\mathbf{a}_i(\mathbf{x},\mathbf{p})$ will be eliminated step by step in the TAYLOR–expansion of $\mathbf{f}$ (equation (2.29)). A necessary and sufficient condition that the process of elimination works is stated in the following theorem:

Theorem (2.1):

If the TAYLOR–expansion of $\mathbf{f}$ starts with polynomials

$$\mathbf{v}_k(\mathbf{x}) := \begin{bmatrix} \mathbf{0} \\ \mathbf{a}_k(\mathbf{x}) \end{bmatrix} \in \Pi_k \ ; \quad k \geq 2 \ ,^5 \tag{2.35}$$

i.e.

$$\dot{\mathbf{x}} = \mathbf{A}\mathbf{x} + \mathbf{v}_k(\mathbf{x}) + \mathbf{v}_{k+1}(\mathbf{x}) + \ldots \ , \tag{2.36}$$

then there is a diffeomorph transformation of the variables

$$\mathbf{x} = \mathbf{z} + \mathbf{P}_k(\mathbf{z}) \ ; \quad \mathbf{P}_k \in \Pi_k \ , \tag{2.37}$$

which transforms the dynamical system (2.34) into the ODE

$$\dot{\mathbf{z}} = \mathbf{A}\mathbf{z} + \mathbf{w}_{k+1}(\mathbf{z}) + \mathbf{w}_{k+2}(\mathbf{z}) + \ldots \tag{2.38}$$

iff the "homological equation"

$$D\mathbf{P}_k(\mathbf{z}) \cdot \mathbf{A}\mathbf{z} - \mathbf{A} \cdot \mathbf{P}_k(\mathbf{z}) = \mathbf{v}_k(\mathbf{z}) \tag{2.39}$$

has a unique solution.

Π_k and $\mathbf{w}_{k+1}$ are defined by

$\Pi_k :=$ set of all vector–polynomials, which consist of monomials of degree $k \geq 2$.

$$\mathbf{w}_{k+1}(\mathbf{z}) = \begin{cases} \mathbf{v}_{k+1}(\mathbf{z}) + D\mathbf{v}_k(\mathbf{z})\,\mathbf{P}_k(\mathbf{z}) & k = 2 \\ \mathbf{v}_{k+1}(\mathbf{z}) & k > 2 \end{cases} \tag{2.40}$$

[5] For the sake of clarity, the parameter $\mathbf{p}$ is not explicitly mentioned.

Proof:

" $\Longleftarrow$ " If the polynomial $\mathbf{P}_k \in \Pi_k$ is the (unique) solution of equation (2.39) and the transformation $\mathbf{x} := \mathbf{z} + \mathbf{P}_k(\mathbf{z})$ is employed in equation (2.36), then equation (2.38) results.

" $\Longrightarrow$ " Define $\mathbf{x} = \mathbf{z} + \mathbf{P}_k(\mathbf{z})$ with $\mathbf{P}_k \in \Pi_k$

$$\Longrightarrow \qquad \dot{\mathbf{x}} = \dot{\mathbf{z}} + D\mathbf{P}_k(\mathbf{z})\dot{\mathbf{z}} \ . \tag{2.41}$$

Define furthermore

$$\dot{\mathbf{z}} = \mathbf{A}\mathbf{z} + \mathbf{w}_{k+1}(\mathbf{z}) + \dots \tag{2.42}$$

with

$$\mathbf{w}_{k+1}(\mathbf{z}) \in \Pi_{k+1} \ .$$

Then on the one hand from equations (2.36) and (2.41) we have:

$$\begin{aligned}
\dot{\mathbf{x}} &= \mathbf{A}\mathbf{x} + \mathbf{v}_k(\mathbf{x}) + \mathbf{v}_{k+1}(\mathbf{x}) + \mathbf{o}_{k+2}(\mathbf{x}) \\
&= \mathbf{A}\left(\mathbf{z} + \mathbf{P}_k(\mathbf{z})\right) + \mathbf{v}_k\left(\mathbf{z} + \mathbf{P}_k(\mathbf{z})\right) + \mathbf{v}_{k+1}\left(\mathbf{z} + \mathbf{P}_k(\mathbf{z})\right) + \mathbf{o}_{k+2}(\mathbf{z}) \\
&= [\mathbf{A}\mathbf{z} + \mathbf{A}\mathbf{P}_k(\mathbf{z})] + [\mathbf{v}_k(\mathbf{z}) + D\mathbf{v}_k(\mathbf{z})\mathbf{P}_k(\mathbf{z}) + \mathbf{o}_{3k-2}(\mathbf{z})] + \\
&\quad + [\mathbf{v}_{k+1}(\mathbf{z}) + \mathbf{o}_{2k}(\mathbf{z})] + \mathbf{o}_{k+2}(\mathbf{z}) \ .
\end{aligned} \tag{2.43}$$

On the other hand from equations (2.41) and (2.42) results:

$$\begin{aligned}
\dot{\mathbf{z}} + D\mathbf{P}_k(\mathbf{z}) \cdot \dot{\mathbf{z}} &= [\mathbf{A}\mathbf{z} + \mathbf{w}_{k+1}(\mathbf{z}) + \mathbf{o}_{k+2}(\mathbf{z})] + D\mathbf{P}_k(\mathbf{z})\,[\mathbf{A}\mathbf{z} + \mathbf{w}_{k+1}(\mathbf{z}) + \\
&\quad + \mathbf{o}_{k+2}(\mathbf{z})] \ , \\
&= \mathbf{A}\mathbf{z} + \mathbf{w}_{k+1}(\mathbf{z}) + D\mathbf{P}_k(\mathbf{z})\mathbf{A}\mathbf{z} + \mathbf{o}_{k+2}(\mathbf{z}) \ .
\end{aligned} \tag{2.44}$$

Comparing polynomials of the same degree equations (2.45) and (2.46) will follow

$$D\mathbf{P}_k(\mathbf{z}) \cdot \mathbf{A}\mathbf{z} - \mathbf{A}\mathbf{P}_k(\mathbf{z}) = \mathbf{v}_k(\mathbf{z}) \tag{2.45}$$

$$\begin{aligned} \mathbf{w}_{k+1}(\mathbf{z}) &= \mathbf{v}_{k+1}(\mathbf{z}) + D\mathbf{v}_k(\mathbf{z}) \cdot \mathbf{P}_k(\mathbf{z}) & k = 2 \ , \\ \mathbf{w}_{k+1}(\mathbf{z}) &= \mathbf{v}_{k+1}(\mathbf{z}) & k > 2 \ . \end{aligned} \tag{2.46}$$

$\Longrightarrow \mathbf{v}_k$ can be eliminated if equation (2.39) has a unique solution $\mathbf{P}_k \in \Pi_k$.

□

An interesting side effect is that $\mathbf{w}_{k+1}(\mathbf{z})$ has the same structure than $\mathbf{v}_k(\mathbf{z})$:

$$\mathbf{w}_{k+1}(\mathbf{z}) = \begin{bmatrix} \mathbf{0} \\ \mathbf{a}_k(\mathbf{z}) \end{bmatrix} + \begin{cases} \begin{pmatrix} \mathbf{0} \\ D\mathbf{a}_k(\mathbf{z})\mathbf{P}_k(\mathbf{z}) \end{pmatrix} & k = 2 \\ \mathbf{0} & k > 2 \end{cases} \ . \tag{2.47}$$

In order to solve the homological equation (2.39) an additional transformation $\mathbf{z} = \mathbf{T}\mathbf{y}$, $\mathbf{T} \in GL(n,n)$ is necessary, which maps $\mathbf{A}$ into diagonalform or JORDAN-normal form. After the multiplication of $\mathbf{T}^{-1}$ with equation (2.39) from the left hand side

$$\mathbf{T}^{-1} D_{\mathbf{z}}\mathbf{P}_k(\mathbf{T}\mathbf{y})\mathbf{T}\left(\mathbf{T}^{-1}\mathbf{A}\mathbf{T}\right)\mathbf{y} - \left(\mathbf{T}^{-1}\mathbf{A}\mathbf{T}\right)\mathbf{T}^{-1}\mathbf{P}_k(\mathbf{T}\mathbf{y}) = \mathbf{T}^{-1}\mathbf{v}_k(\mathbf{T}\mathbf{y}) \tag{2.48}$$

results.

Taking the chain rule

$$\begin{aligned} D_{\mathbf{z}}\mathbf{P}_k(\mathbf{T}\mathbf{y})\mathbf{T} &= D_{\mathbf{z}}\mathbf{P}_k(\mathbf{T}\mathbf{y}) \cdot D_{\mathbf{y}}\mathbf{z} \\ &= D_{\mathbf{y}}\left(\mathbf{P}_k \circ \mathbf{T}\right)(\mathbf{y}) \end{aligned} \tag{2.49}$$

into account, equation (2.48) together with the abbreviations

$$\begin{aligned} \mathbf{R}_k &:= \mathbf{T}^{-1}\mathbf{P}_k \circ \mathbf{T} & \in \Pi_k \ , \\ \mathbf{J} &:= \mathbf{T}^{-1}\mathbf{A}\mathbf{T} & \in \mathbb{R}^{n,n} \ , \\ \mathbf{b}_k &:= \mathbf{T}^{-1}\mathbf{v}_k \circ \mathbf{T} & \in \Pi_k \end{aligned} \tag{2.50}$$

leads to the simpler form

$$D\mathbf{R}_k(\mathbf{y}) \cdot \mathbf{J}\mathbf{y} - \mathbf{J} \cdot \mathbf{R}_k(\mathbf{y}) = \mathbf{b}_k(\mathbf{y}) \ . \tag{2.51}$$

If $\mathbf{J}$ has diagonalform (i.e. $\mathbf{J} = diag\{\lambda_1, \ldots, \lambda_n\}$), theorem (2.2) is valid.

Theorem (2.2):

If $\Pi := \bigoplus_{k\in\mathbb{N}} \Pi_k$, then

$$\mathbf{L_J} : \Pi \to \Pi \ , \quad \mathbf{P}(\mathbf{y}) \mapsto D\mathbf{P}(\mathbf{y})\mathbf{J}\mathbf{y} - \mathbf{J}\cdot\mathbf{P}(\mathbf{y}) \tag{2.52}$$

defines a linear operator with the

$$\begin{array}{ll} (i) & \text{eigenvalues } \sigma_{m,j} := <\mathbf{m}, \boldsymbol{\lambda}> -\lambda_j \\ & \text{and the} \\ (ii) & \text{eigenvectors } \mathbf{P}_{m,j}(\mathbf{y}) := \mathbf{y}^{\mathbf{m}}\mathbf{e}_j \ . \end{array} \tag{2.53}$$

$\mathbf{y}^{\mathbf{m}} := y_1^{m_1} \cdot \ldots \cdot y_n^{m_n}$ defines a monomial of degree $|\,\mathbf{m}\,| \in \mathbb{N}$. $m_i \in \mathbb{N}$ and $j \in \mathbb{N}$ are arbitrary integers.

Proof: Employ equations (2.53) and (2.52).

□

According to theorem (2.2) the operator equation

$$\mathbf{L_J}(\mathbf{R}_k) = \mathbf{b}_k \tag{2.54}$$

has a unique solution $\mathbf{R}_k \in \Pi_k$ iff the eigenvalues $\sigma_{m,j}$ ($|\,\mathbf{m}\,| = k$ and $j \in \{1, \ldots, n\}$) do not disappear.

Namely, if

$$\mathbf{b}_k(\mathbf{y}) = \sum_{\substack{\mu,\nu \\ |\mu| = k}} \beta_{\mu,\nu}\mathbf{y}^{\mu}\mathbf{e}_{\nu} \ , \tag{2.55}$$

then theorem (2.2) says

$$\mathbf{R}_k(\mathbf{y}) = \sum_{\substack{\mu,\nu \\ |\mu| = k}} r_{\mu,\nu}\mathbf{y}^{\mu}\mathbf{e}_{\nu} \ , \tag{2.56}$$

where $r_{\mu,\nu}$ is defined by

$$r_{\mu,\nu} := \frac{\beta_{\mu,\nu}}{\sigma_{\mu,\nu}} \ . \tag{2.57}$$

Monomials $\mathbf{y}^{\mathbf{m}}\mathbf{e}_j$ in the series expansion of $\mathbf{f}$ are called "resonant" if the eigenvalues $\sigma_{m,j}$ belonging to this monomial are zero-eigenvalues. Resonant monomials cannot be eliminated by the method mentioned above.

For non-resonant monomials, the process of elimination may be carried out qualitatively as follows:

$\mathbf{x}^{(1)} := \mathbf{x} \in \mathbb{R}^n$

$\mathbf{v}_2^{(1)} := \mathbf{v}_2$

Determine $\mathbf{T} \in GL(n,n) \quad \Rightarrow \quad \mathbf{J} = \mathbf{T}^{-1}\mathbf{A}\mathbf{T}$

$\underline{k = 2,3,4,\ldots}$

$$
\begin{aligned}
\mathbf{b}_k(\mathbf{y}) &:= \left(\mathbf{T}^{-1}\mathbf{v}_k^{k-1} \circ \mathbf{T}\right)(\mathbf{y}) \\
&= \sum_{\substack{\mu,\nu \\ |\mu| = k}} \beta_{\mu,\nu}^{(k-1)} \mathbf{y}^{\mu} \mathbf{e}_{\nu} \\
\mathbf{R}_k(\mathbf{y}) &:= \sum_{\substack{\mu,\nu \\ |\mu| = k}} \left(\frac{\beta_{\mu,\nu}^{(k-1)}}{\sigma_{\mu,\nu}}\right) \mathbf{y}^{\mu} \mathbf{e}_{\nu} \\
\mathbf{P}_k(\mathbf{y}) &:= \left(\mathbf{T}^{-1}\mathbf{R}_k \circ \mathbf{T}\right)(\mathbf{y}) \\
\mathbf{x}^{(k)} &:= \mathbf{x}^{(k+1)} + \mathbf{P}_k\left(\mathbf{x}^{(k+1)}\right) \\
\mathbf{v}_{k+1}^{(k)} &:= \mathbf{v}_{k+1}^{(k-1)}
\end{aligned}
\tag{2.58}
$$

The main difficulty in applying the algorithm (2.58) is to compute the terms

$$
\mathbf{v}_k^{(k-1)},\ \mathbf{v}_{k+1}^{(k-1)},\ \mathbf{v}_{k+2}^{(k-1)},\ldots \qquad (k = 2,3,4,\ldots)\ . \tag{2.59}
$$

However, for practical investigations it may be useful to eliminate the monomials up to degree n $(n = 2,3,4,\ldots)$. Since the most essential properties of the dynamical behavior are determined by the first terms of the series expansion of $\mathbf{f}$.

Equations (2.50) and (2.51) lead to

$$
\mathbf{L}_{\mathbf{T}^{-1}\mathbf{A}\mathbf{T}} = \mathbf{T}^{-1}\mathbf{L}_{\mathbf{A}} \circ \mathbf{T} \tag{2.60}
$$

and therefore

$$
\mathbf{L}_{\mathbf{J}} = \mathbf{T}^{-1}\mathbf{L}_{\mathbf{A}} \circ \mathbf{T}\ . \tag{2.61}
$$

That means that $\mathbf{L}_{\mathbf{A}}$ is transformed in the same way than $\mathbf{A}$ is. Hence, if $\mathbf{J}$ has diagonalform then $\mathbf{L}_{\mathbf{J}}$ is "diagonal" too. If $\mathbf{J}$ has JORDAN–normal form, $\mathbf{L}_{\mathbf{J}}$ has also JORDAN–normal form and the eigenvalues $\sigma_{m,j}$ are the same too. That means

$\mathbf{L_J}$ is also invertible, if $\mathbf{A}$ has multiple eigenvalues, provided that there are no zero-eigenvalues $\sigma_{m,j}$.

∇

Example 2.3: (damped double pendulum)

The equations of motion of a double pendulum which is damped proportionally to its velocity (for instance air friction and bearing friction) and moving in the gravitational field of the earth are:

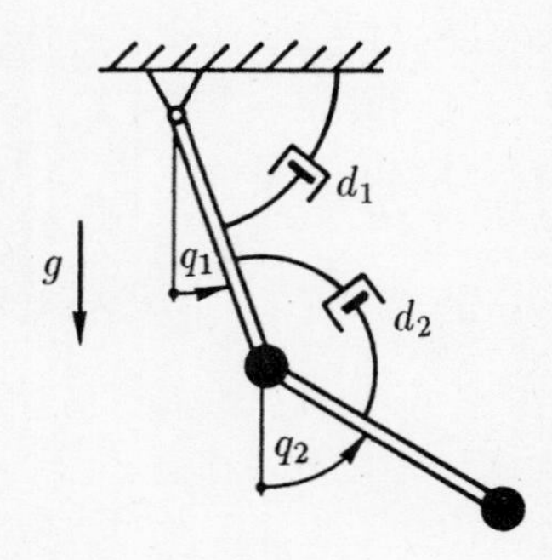

Fig. 2.2: double pendulum

$$\begin{aligned}2\ddot{q}_1 + \cos(q_1 - q_2)\,\ddot{q}_2 + \dot{q}_2^2 \sin(q_1 - q_2) + 2\omega_0^2 \sin(q_1) + \varepsilon_1 \dot{q}_1 - \varepsilon_2 \dot{q}_2 &= 0\\ \cos(q_1 - q_2)\,\ddot{q}_1 + \ddot{q}_2 - \dot{q}_1^2 \sin(q_1 - q_2) + \omega_0^2 \sin(q_2) - \varepsilon_2 \dot{q}_1 + \varepsilon_2 \dot{q}_2 &= 0\end{aligned}$$

$$\text{where} \quad \omega_0^2 := \frac{g}{l} \quad ; \quad \varepsilon_1 := \frac{d_1 + d_2}{ml^2} \quad ; \quad \varepsilon_2 := \frac{d_2}{ml^2} \ .$$

Here, the masses and lengths of the two pendulums are to be assumed equal. The linearized equation of motion around the fix point $\mathbf{x}_0 = \left(\mathbf{q}_0^T, \dot{\mathbf{q}}_0^T\right)$ is

$$\begin{bmatrix} 2 & -1 \\ -1 & 1 \end{bmatrix} \ddot{\overline{\mathbf{q}}} + \begin{bmatrix} \varepsilon_1 & -\varepsilon_2 \\ -\varepsilon_2 & \varepsilon_2 \end{bmatrix} \dot{\overline{\mathbf{q}}} + \begin{bmatrix} 2\omega_0^2 & 0 \\ 0 & -\omega_0^2 \end{bmatrix} \overline{\mathbf{q}} = 0$$

where $\mathbf{q} = \mathbf{q}_0 + \overline{\mathbf{q}}$; $\| \overline{\mathbf{q}} \| \ll 1$.

The eigenvalues of this linear system are

$$\begin{aligned}\lambda_1^{\pm} &= \delta_1(\boldsymbol{\varepsilon}) \pm i\omega_1(\boldsymbol{\varepsilon}) \ ,\\ \lambda_2^{\pm} &= \delta_2(\boldsymbol{\varepsilon}) \pm i\omega_2(\boldsymbol{\varepsilon}) \ ,\\ \text{where} \qquad \boldsymbol{\varepsilon} &:= (\varepsilon_1, \varepsilon_2)^T\\ \text{and} \qquad \delta_i(\boldsymbol{\varepsilon} = \mathbf{0}) &= 0 \ ; \quad (i = 1, 2) \ ,\\ \omega_1(\boldsymbol{\varepsilon} = \mathbf{0}) &= \sqrt{\tfrac{\sqrt{2}}{\sqrt{2}-1}}\,\omega_0 \ ,\\ \omega_2(\boldsymbol{\varepsilon} = \mathbf{0}) &= \sqrt{\tfrac{\sqrt{2}}{\sqrt{2}+1}}\,\omega_0 \ .\end{aligned}$$

$\varepsilon_i > 0$ means $\delta_i > 0$ $(i = 1,2)$. According to equation (2.53) the eigenvalues $\sigma_{m,j}$ are not resonant if $\delta_i > 0$, but they are resonant if $\delta_i = 0$ is the case $(i\,\omega_j = m_1\,(i\omega_j) + m_2\,(-i\omega_j)\;\forall\, m_1 = m_2 + 1!)$.

That means if the system is damped $(\varepsilon_i > 0)$, then it is diffeomorph to its linear part obtained by linearization around the fix point.

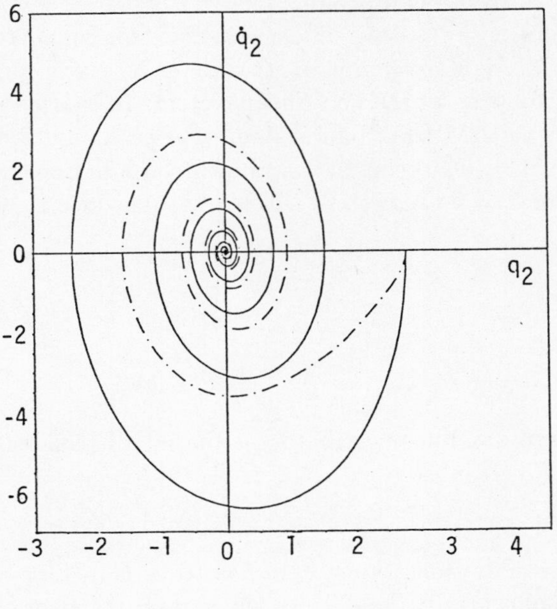

—— *linearized system*
- · - *non-linear system*

Fig. 2.3: phase space trajectory $(q, \dot{q})$ of the damped double pendulum

In other words: For each solution of the non-linear system there is a solution of the equivalent linear system and a transformation

$$\mathbf{P} := \prod_{k=2}^{\infty} (\mathbf{E} + \mathbf{P}_k) = (\mathbf{E} + \mathbf{P}_2) \circ (\mathbf{E} + \mathbf{P}_3) \circ \ldots \;, \tag{2.62}$$

which may be constructed according to equation (2.58). **P** transforms solutions of the linear system diffeomorphically into their equivalent solutions of the non-linear system.

This is not the case for singular (unstable) solutions

$$\mathbf{q}(t) = (0\,,\,\pi)$$

or

$$\mathbf{q}(t) = (\pi\,,\,0)$$

or

$$\mathbf{q}(t) = (\pi\,,\,\pi)\;,$$

respectively which denote the unstable points of equilibrium of the system. These points can not be transformed by **P** into a solution of the linearized system around $\mathbf{q}(t) = (0\,,\,0)$.

From the geometrical point for view, the reason for the existence of the transformation **P** is that all solutions of the non-linear damped systems are connected at the (stable) point $\mathbf{q} = (0\,,\,0)$ of equilibrium. That means there is only one connected component of the solutions of the non-linear system. This connected component will decompose into infinitely many components if $\delta_i \to 0$. Each component will then be represented by a solution of the non-linear system. In that case (i.e. $\delta_i \to 0$) the system converges to a HAMILTONian system, and each solution of the linearized system around point $\mathbf{q} = (0\,,\,0)$ consists of a linear combination of periodic solutions (eigen-solutions). For $\delta_i = 0$ the system will be resonant (see chapter 2.5).

△

▽

Example 2.4:

One of the well known non-linear oscillators is the DUFFING oscillator:

$$\ddot{x} + \alpha\dot{x} + \beta x^3 = \gamma \cos \omega t \ , \qquad (\alpha, \beta, \gamma > 0) \ .$$

For the following considerations the system has to be transformed into an autonomous ODE. This may be easily done, since the system is harmonically excited.

$$\left.\begin{aligned} \ddot{y} + \omega^2 y &= 0 \\ y(0) &= \gamma \\ \dot{y}(0) &= 0 \end{aligned}\right\} \iff \quad y(t) = \gamma \cos \omega t \ .$$

Hence, the equivalent autonomous system is

$$\begin{aligned} \ddot{x} + \alpha\dot{x} + \beta x^3 - y &= 0 \ , \\ \ddot{y} + \omega^2 y &= 0 \end{aligned}$$

or if transformed into a system of first order $\left(\mathbf{x}^T := (x_1, x_2, x_3, x_4) := (x, y, \dot{x}, \dot{y})\right)$

$$\dot{\mathbf{x}} = \underbrace{\begin{bmatrix} 0 & 0 & 1 & 0 \\ 0 & 0 & 0 & 1 \\ 0 & 1 & -\alpha & 0 \\ 0 & -\omega^2 & 0 & 0 \end{bmatrix}}_{=:\mathbf{A}^*} \mathbf{x} + \underbrace{\begin{bmatrix} 0 \\ 0 \\ -\beta x_1^3 \\ 0 \end{bmatrix}}_{=:\mathbf{v}_3(x)}$$

respectively.

$\mathbf{A}^*$ has the eigenvalues $\lambda_1 = 0$, $\lambda_2 = -\alpha$, $\lambda_3 = i\omega$, $\lambda_4 = -i\omega$. According to equation (2.53) the eigenvalues λ_1, λ_3 and λ_4 are resonant. Therefore, the non-linear system can not be transformed into its linearized system around the fixed point $\mathbf{x} = \mathbf{0}$.

However, if a convenient "state feedback" or a "singular disturbance" respectively is introduced, a vector field with no resonances can be obtained:

$$\begin{aligned} \dot{\mathbf{x}} &= \mathbf{f}^*(\mathbf{x}, \mathbf{u}) \ , \\ \mathbf{0} &= \mathbf{g}(\mathbf{x}, \mathbf{u}) \end{aligned}$$

where

$$\begin{aligned} \mathbf{f}^*(\mathbf{x}, \mathbf{u}) &:= \mathbf{A}^*\mathbf{x} + \mathbf{v}_3(\mathbf{x}) + \mathbf{B}\mathbf{u} \ , \\ \mathbf{B} &:= (0 \ , \ 0 \ , \ 0 \ , \ 1)^T \end{aligned}$$

and

$$\mathbf{g}(\mathbf{x}, \mathbf{u}, \varepsilon) := u - \varepsilon x_1$$

leads – according to equation (2.27) – after elimination of u to the dynamical system:

$$\dot{\mathbf{x}} = \mathbf{f}(\mathbf{x}, \boldsymbol{\varepsilon})$$

with

$$\mathbf{f}(\mathbf{x}, \mathbf{p}) := \underbrace{\begin{bmatrix} 0 & 0 & 1 & 0 \\ 0 & 0 & 0 & 1 \\ 0 & 1 & -\alpha & 0 \\ \varepsilon & -\omega^2 & 0 & 0 \end{bmatrix}}_{=:\mathbf{A}} \mathbf{x} + \begin{bmatrix} 0 \\ 0 \\ -\beta x_1^3 \\ 0 \end{bmatrix} .$$

The vector $\mathbf{p} := (\alpha, \ \beta, \ \omega, \ \varepsilon)$ denotes the parameter of the system.

The fixed point $\mathbf{x} = \mathbf{0}$ which is triple and degenerated will be "universally" unfolded into three conventional fixed points (c.f. example 2.7):

$$\mathbf{x}^{(1)} := \begin{bmatrix} -\sqrt{\frac{\varepsilon}{\omega^2\beta}} \\ -\frac{\varepsilon}{\omega^2}\sqrt{\frac{\varepsilon}{\omega^2\beta}} \\ 0 \\ 0 \end{bmatrix} \ ; \quad \mathbf{x}^{(2)} := 0 \ ; \quad \mathbf{x}^{(3)} := \begin{bmatrix} +\sqrt{\frac{\varepsilon}{\omega^2\beta}} \\ -\frac{\varepsilon}{\omega^2}\sqrt{\frac{\varepsilon}{\omega^2\beta}} \\ 0 \\ 0 \end{bmatrix} .$$

The eigenvalues of the modified matrix **A** are determined by the equation

$$\begin{aligned} 0 &= det\,(\mathbf{A} - \lambda\mathbf{E}) \;, \\ &= det\,(\mathbf{A} - \lambda\mathbf{E}) - \varepsilon \;, \\ &= \lambda(\alpha + \lambda)\,(\lambda^2 + \omega^2) - \varepsilon \;, \\ &= (\lambda - \lambda_1^*)\,(\lambda - \lambda_2^*)\,(\lambda - \lambda_3^*)\,(\lambda - \lambda_4^*) - \varepsilon \;, \end{aligned}$$

where $\lambda_1^*, \ldots, \lambda_4^*$ denote the eigenvalues of the matrix $\mathbf{A}^*$.

The modification $p^*(\lambda) := p(\lambda) - \varepsilon$ of the graph of $p(\lambda) := det(\mathbf{A} - \lambda\mathbf{E})$ causes a vertical shift.

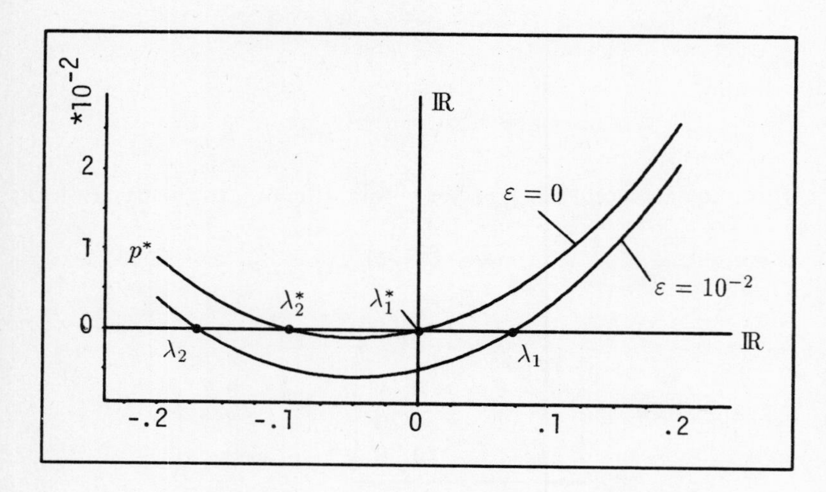

Fig. 2.4: characteristic polynomial p or p^* respectively

The eigenvalues $\lambda_1^*, \ldots, \lambda_4^*$ are modified and the system loses its stability. For $\varepsilon < 0$ the fix point $\mathbf{x} = \mathbf{0}$ is stable.

As stated in equation (2.58) the now non-resonant vector field may be transformed into its normal form around the fix point. The first step, doing that is to transform **A** into its diagonalform with a non-singular matrix $\mathbf{T} \in \mathbb{R}^{4,4}$. The columns of **T** are the eigenvectors of **A** which has the eigenvalues λ_i. The eigenvalues and also the eigenvectors have to be computed numerically. Only the polynomial $\mathbf{v}_3(\mathbf{x}) = -\beta x_1^3 \mathbf{e}_3$ needs to be eliminated. If the coefficients $\alpha, \boldsymbol{\mu}, \nu$ are determined by the equation

$$\begin{aligned} \mathbf{b}_3(\mathbf{y}) &:= \mathbf{T}^{-1}\mathbf{v}_3(\mathbf{T}\mathbf{y}) \;, \\ &= \sum_{\substack{\mu,\nu \\ |\mu|=3}} \beta_{\mu,\nu}\, \mathbf{y}^{\mu} \end{aligned}$$

the desired transformation is obtained by:

$$\mathbf{x} = \mathbf{z} + \mathbf{P}_3(\mathbf{z})$$

where

$$\mathbf{P}_3(\mathbf{z}) := \mathbf{T} \cdot \mathbf{R}_3(\mathbf{T}^{-1}\mathbf{z}) ,$$

$$\mathbf{R}_3(\mathbf{y}) := \sum_{\substack{\mu,\nu \\ |\mu|=3}} \left(\frac{\beta_{\mu,\nu}}{\sigma_{\mu,\nu}}\right) \mathbf{y}^{\mu} .$$

This transformation maps the trajectories of the non-linear system into the trajectories of the linearized system around the fixed point $\mathbf{x} = \mathbf{0}$.

△

Regarding the transformation into normal form, the following question arises: For what set of points $\mathbf{x} \in TM$ does the series, constructed according to equation (2.58), converge?

If, for instance, $\mathbf{x} = \mathbf{0}$ is a fixed point of the non-resonant vector field $\mathbf{f}$, then (in an open neighborhood of that fixed point) a transformation $\mathbf{z} = \mathbf{Q}(\mathbf{x})$ into normal form exists:

$$\begin{array}{ccc} TM & \xrightarrow{\mathbf{f}} & TM \\ Q \downarrow & & \downarrow DQ(\cdot) \\ T_0M & \xrightarrow{D\mathbf{f}(0)} & T_0M \end{array}$$

For an arbitrary point $\mathbf{x}$ in a neighborhood of $\mathbf{0}$ we have

$$D\mathbf{f}(\mathbf{0}) \cdot Q(\mathbf{x}) = D\,Q(\mathbf{x}) \cdot \mathbf{f}(\mathbf{x}) . \tag{2.63}$$

If $\mathbf{x} \neq \mathbf{0}$ is a different singular point, then $\mathbf{f}(\mathbf{x}) = \mathbf{0}$.

$$\Rightarrow \qquad D\mathbf{f}(\mathbf{0})\, Q(\mathbf{x}) = \mathbf{0} . \tag{2.64}$$

$D\mathbf{f}(\mathbf{0})$ is regular (zero-eigenvalues are resonant and therefore they do not exist)

$$\Rightarrow \qquad Q(\mathbf{x}) = \mathbf{0} .$$

That is a contradiction to the assumption that $\mathbf{Q}$ is bijective, since $\mathbf{x} = \mathbf{0}$ as well as $\mathbf{x} \neq \mathbf{0}$ are zeros of $\mathbf{Q}$. That means each area of convergence may only contain one singularity. Furthermore, if the singularity is unstable then one or more separatrices intersect the singularity. The separatrices are manifolds, which separate the distinct areas of convergence of the transformation into normal form. And each area of convergence contains one and only one singularity.

Another important aspect, not mentioned so far, is the behavior of the vector field $\mathbf{f}$ as a map, if the value of the parameter $\mathbf{p}$ is fixed. If $\mathbf{x}_0 \in TM$ is a fixed point of $\mathbf{f}(\cdot\,, \mathbf{p}) : TM \to TM$, then it turns out – like in theorem (2.1) – (c.f. [ARNOLD 1983]) that a diffeomorph transformation $\mathbf{z} = \mathbf{Q}(\mathbf{x})$ exists, which maps the field $\mathbf{x} \mapsto \mathbf{f}(\mathbf{x}, \mathbf{p})$ into the linearized field $\mathbf{z} \mapsto D\mathbf{f}(\mathbf{x}_0, \mathbf{p})\mathbf{z}$ around the fixed point $\mathbf{x}_0$. As above, the eigenvalues λ_i of $D\mathbf{f}(\mathbf{x}_0, \mathbf{p})$ have to be non-resonant. The condition of resonance is given by

$$\lambda_k = \lambda_1^{m_1} \cdot \ldots \cdot \lambda_n^{m_n} = \boldsymbol{\lambda}^{\mathbf{m}} \,, \tag{2.65}$$

where $k \in \{1, \ldots, n\}$, $\mathbf{m} \in \mathbb{N}^n$ and $|\,\mathbf{m}\,| \geq 2$.

2.4 Classification of singularities

[POINCARÉ 1893] classified the singularities of a two dimensional vector field **f** according to the dynamical behavior of **f** around their singularities (i.e. zeros or points of equilibrium of the field **f**). He assumed that the flow of the non-linear dynamical system in the neighborhood of the singularity is homeomorph to the flow of the linearized system around this fixed point. That means in the neighborhood of the singularity the non-linear system and its linear part are topologically equivalent. If that is not the case then the singularity is called "degenerated". A systematic classification of degenerated singularities can be found in [ARNOLD 1985]. Unfortunately the discussion about the singularities of vector fields is mostly very general. Their importance and consequence for dynamical systems is usually not mentioned.

For analytic vector fields the condition of nonresonancy

$$\sigma_{m,j} = < \mathbf{m}, \boldsymbol{\lambda} > - \lambda_j \neq 0 \qquad (2.66)$$

is a sufficient criterion, since in that case the necessary homeomorph transformation can be immediately constructed. Zero-eigenvalues ($\lambda_j = 0$), pure imaginary pairs of eigenvalues ($\lambda_{j,j+1} = \pm i\omega_j$) as well as the representation of an eigenvalue by a linearcombination of eigenvalues ($\lambda_j = \sum_{k=1}^{n} c_k \lambda_k$) are resonant. The theorem of GROBMANN and HARTMAN proves the topological equivalence between the non-linear system and its linear part, if there are no zero-eigenvalues or pure imaginary pairs of eigenvalues. That means, equation (2.66) is not a necessary condition. Pairs of pure imaginary eigenvalues lead in linear systems always to periodic solutions. The dynamical behavior in linear subspaces in which periodic solutions of the linearized system exist is topologically equivalent to a HAMILTONian or conservative (sub-)system of the non-linear dynamical system ([WEINSTEIN 1973], [MOSER 1976]).

Zero-eigenvalues lead in linear systems necessarily to constant solutions, that means to points in the state space, which are of course not homeomorph to time-variant solutions of the non-linear system in the neighborhood of the singularity. Therefore a singularity is called "degenerated into the direction of an eigenvector" if the eigenvalue to that eigenvector is a zero-eigenvalue. Otherwise, the singularity is called regular.

The classification of regular singularities of smooth vector fields or the dynamical behavior of the system around the singularity respectively can be investigated by means of the "center manifold theorem", which is a generalization of POINCARÉ's classical classification of systems with one DOF.

The first step of that method of investigation is to separate the eigenvalues of the JACOBIan $D\mathbf{f}(\mathbf{0})$ of the vector field **f** at the singular point $\mathbf{x} = \mathbf{0}$ into three groups:

$$
\begin{aligned}
\Lambda_s &:= \{\lambda_j \mid Re\,\lambda_j < 0\} \ , \\
\Lambda_c &:= \{\lambda_j \mid Re\,\lambda_j = 0\} \ , \\
\Lambda_u &:= \{\lambda_j \mid Re\,\lambda_j > 0\} \ .
\end{aligned}
\tag{2.67}
$$

If E_s, E_c or E_u respectively are the eigenspaces, which belong to Λ_s, Λ_c or Λ_u respectively, then the center manifold theorem says: There are invariant manifolds M_s, M_c and M_u with $T_0M_s = E_s$; $T_0M_c = E_c$ and $T_0M_u = E_u$ in the neighborhood of the singularity $\mathbf{x} = \mathbf{0}$ of the vector field $\mathbf{f}$. M_s and M_u have a unique representation.

∇

Example 2.5:

The dynamical system

$$
\begin{bmatrix} \dot{x}_1 \\ \dot{x}_2 \end{bmatrix} = \begin{bmatrix} -x_1\,(x_2+1) \\ (x_1^2-1)\,(x_1^2+x_2^2) \end{bmatrix}
$$

has three fix points, two regular points $\mathbf{x} = (\pm 1,\ -1)$ and one degenerated point $\mathbf{x} = \mathbf{0}$. $\mathbf{x} = \mathbf{0}$ is a crossing point of a stable manifold M_s and a center manifold M_c.

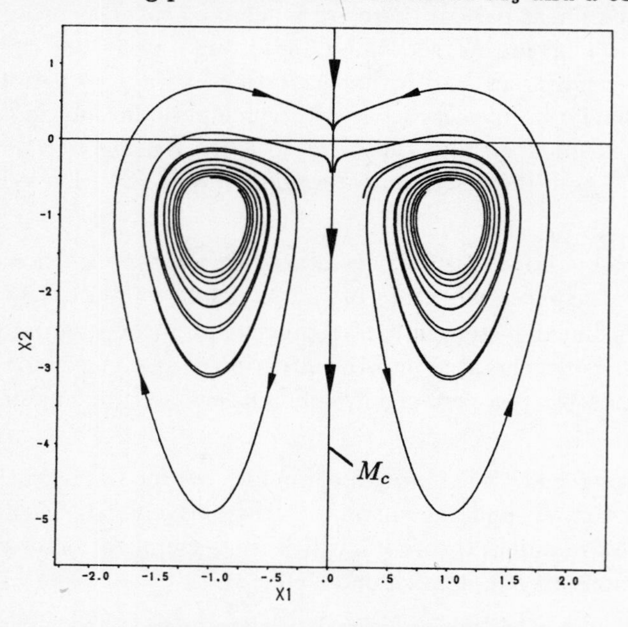

Fig. 2.5: center manifold M_c

Δ

The expression "invariant manifold" means that a trajectory starting on that manifold will stay on it for ever. A trajectory starting on M_s moves to the singularity (stable manifold), a trajectory starting on M_u leaves the singularity (unstable manifold). A trajectory which starts on $M_s \times M_u$ will be unstable too. The flow on M_c depends on further investigations, since E_c contains all degenerated directions.

2.4.1 Corank, critical and noncritical variables

In the neighborhood of the singularity $\mathbf{x} = \mathbf{0}$, the vector field $\mathbf{f}$ has the TAYLOR-expansion

$$f_k(\mathbf{x}, \mathbf{p}) = Df_k(\mathbf{0}, \mathbf{p})\,\mathbf{x} + \frac{1}{2}\mathbf{x}^T D^2 f_k(\mathbf{0}, \mathbf{p})\,\mathbf{x} + \ldots \ .$$

$D^2 f_k(\mathbf{0}, \mathbf{p}) \in \mathrm{I\!R}^{2f,2f}$ is the HESSEian of f_k. If $D^2 f_k(\mathbf{0}, \mathbf{p})$ has the rank r then the rank defect $2f - r$ is called "corank of the funktion f_k".

Furthermore, if the singularity $\mathbf{x} = \mathbf{0}$ is degenerated in the direction of the component f_k – i.e. $Df_k(\mathbf{0}, \mathbf{p}) = \mathbf{0}$ – then the theorem of SYLVESTER says that a linear transformation of the coordinates $S : \mathbf{y} \mapsto \mathbf{x}$ exists so that f_k expressed in these coordinates has the representation

$$\begin{aligned}
\tilde{f}_k(\mathbf{y}, \mathbf{p}) &= (f_k \circ S)(\mathbf{y}) \\
&= \mathbf{y}^T \begin{bmatrix} \mathbf{E}_\rho & & 0 \\ & -\mathbf{E}_\tau & \\ 0 & & 0 \end{bmatrix} \mathbf{y} \ + \ \text{terms of higher degree} \\
&= y_1^2 + \ldots + y_\rho^2 - y_{\rho+1}^2 - \ldots - y_{\rho+\tau}^2 \ + \ \text{terms of higher degree}
\end{aligned} \qquad (2.68)$$

$r := \rho + \tau$ is the rank of the HESSEian of f_k. $\mathbf{E}_\rho$ and $\mathbf{E}_\tau$ are unit matrices of the dimension ρ and τ.

The number $\mid \rho \mid$ of positive signs and the number $\mid \tau \mid$ of negative signs determine the type of the surface around the singularity described by the variables $y_1, \ldots, y_{\rho+\tau}$ of the function f_k. The behavior of the surface into the direction $y_{\rho+\tau+1}, \ldots, y_n$ is unknown. To investigate these directions, terms of power greater than one must be taken into account, the same is true for functions depending on only one variable. The theorem of SYLVESTER guarantees the splitting of the variables into two parts: into "critical variables" $y_{\rho+\tau+1}, \ldots, y_n$ and into "noncritical variables" $y_1, \ldots, y_{\rho+\tau}$. The number of the critical variables is an essential criterion to classify the singularity. That number is exactly the corank of the function f_k.

2.4.2 Codimension, specification of the singularities

Basic tools to describe dynamical systems are in particular geometrical objects such as the parameter space $P \subset \mathrm{I\!R}^k$ or the configuration space M of the dimension f, in which each point corresponds to a position of the dynamical system.

In general all these objects have the property to be embedded in a higher dimensional space. The difference between the dimension of the embedding space and the dimension of the geometrical object is defined as "codimension" of the geometrical object.

▽

Example 2.6:

The configuration space $T^2 = S^1 \times S^1$ of the double pendulum which has the lengths l_1 and l_2 is described by the following embedding:

$$T^2 \to \mathbb{R}^3 \ , \quad \begin{bmatrix} q_1 \\ q_2 \end{bmatrix} \longmapsto \begin{bmatrix} x \\ y \\ z \end{bmatrix} := \begin{bmatrix} (l_1 + l_2 \cos q_2) \ \cos q_1 \\ (l_1 + l_2 \cos q_2) \ \sin q_1 \\ l_2 \sin q_2 \end{bmatrix} .$$

I.e. the configuration space T^2 has the codimension 1 or: *codim* $(T^2) = 1$.

△

In order to investigate degenerated singularities it may be necessary to use "geometrical objects" which are very abstract. The determination of these geometrical objects is the most important task. For practical aspects it is useful to emphasize more the idea and the method than the mathematical background. Without any restrictions it can be assumed that $\mathbf{A} := D\mathbf{f}\,(\mathbf{0}, \mathbf{p})$ has diagonal form or JORDAN form respectively. Otherwise a coordinate transformation $\mathbf{x} \mapsto \mathbf{y}$ has to be made as shown in equation (2.48). Furthermore it may be assumed that all zero-eigenvalues of A are placed in front:

$$diag\,\{0\,,\ \ldots\,,\, 0\,,\ \lambda_{s+1}\,,\ \ldots\,,\ \lambda_n\} \ . \tag{2.69}$$

Then the degenerated directions of the singularity of the vector field $\mathbf{f}$ are described by their components $f_1, \ldots, f_s$.

The eigenvalues depend on the parameter $\mathbf{p}$ of the system, hence, the degenerated directions also depend on $\mathbf{p}$. Therefore, if the investigation includes the variation of $\mathbf{p}$, it is necessary to split the tangent space $T_{\mathbf{p}}P$ of the parameter space P at the point $\mathbf{p}$ into the kernel and the image of the linear mapping

$$L_k : T_{\mathbf{p}}P \to \mathbb{C} \ , \quad \mathbf{v} \mapsto < grad\ \lambda_k(\mathbf{p})\ ,\ \mathbf{v} > \ , \tag{2.70}$$

Since a modification of the parameter $\mathbf{p}$ into a direction $\mathbf{w} \in Ker\ L_k = \{\mathbf{v} \in T_{\mathbf{p}}P \mid < grad\,\lambda_k(\mathbf{p}), \mathbf{v} >= 0\}$ does not influence the eigenvalue λ_k locally.

∇

Example 2.7:

In coordinates of the state space the ODE of the oscillator of example 2.4 is

$$\dot{\mathbf{z}} = \mathbf{A}\mathbf{z} + \mathbf{v}_3(\mathbf{z})$$

where

$$\mathbf{A} = \begin{bmatrix} 0 & 0 & 1 & 0 \\ 0 & 0 & 0 & 1 \\ 0 & 1 & -\alpha & 0 \\ 0 & -\omega^2 & 0 & 0 \end{bmatrix} \quad \text{and} \quad \mathbf{v}_3(\mathbf{z}) = \begin{bmatrix} 0 \\ 0 \\ -\beta z_1^3 \\ 0 \end{bmatrix} .$$

If $\mathbf{z} = \mathbf{T}\mathbf{x}$ transforms $\mathbf{A}$ into its diagonal form, then we get the form

$$\begin{aligned} \dot{\mathbf{x}} &= (\mathbf{T}^{-1}\mathbf{A}\mathbf{T})\,\mathbf{x} + \mathbf{T}^{-1}\mathbf{v}_3(\mathbf{T}\mathbf{x}) \ , \\ &= \underbrace{\begin{bmatrix} 0 & 0 & 0 & 0 \\ 0 & -\alpha & 0 & 0 \\ 0 & 0 & i\omega & 0 \\ 0 & 0 & 0 & -i\omega \end{bmatrix}}_{=:\boldsymbol{\Lambda}} \mathbf{x} + \mathbf{T}^{-1} \begin{bmatrix} 0 \\ 0 \\ -\beta\left(\sum_{j=1}^4 t_{1j}x_j\right) \\ 0 \end{bmatrix} . \end{aligned}$$

The elements t_{ij} of the matrix $\mathbf{T}$ or τ_{ij} of $\mathbf{T}^{-1}$ respectively depend on the parameters α and ω. Variable parameters are $\mathbf{p} := (\alpha, \beta, \omega)$.

$$\Rightarrow \ \mathbf{f}(\mathbf{x}, \mathbf{p}) := \boldsymbol{\Lambda}\mathbf{x} - \beta \left(\sum_{j=1}^{4} t_{1j}x_j \right)^3 \begin{bmatrix} \tau_{13} \\ \tau_{23} \\ \tau_{33} \\ \tau_{43} \end{bmatrix} .$$

For $\alpha \neq 0$ and $\beta \neq 0$, f_1 is the only degenerated direction. The direction f_2 might be degenerated for $\alpha = 0$.

Furthermore, we have

$$
\begin{array}{llcl}
(1) & grad\,\lambda_1(\mathbf{p}) & = & 0 \qquad \forall\ \mathbf{p} \in P = \mathbb{R}^2 \times \mathbb{R}_+ \ . \\
 & \Rightarrow\ Ker\,L_1 & = & \mathbb{R}^3 \left(\text{or } \mathbb{C}^3\right) \quad \forall\ \mathbf{p} \in P \ . \\
(2) & grad\,\lambda_2(\mathbf{p}) & = & (-1\,,\,0\,,\,0)\ . \\
 & \Rightarrow\ Ker\,L_2 & \cong & \mathbb{R}^2 \left(\text{or } \mathbb{C}^2\right) \quad \forall\ \mathbf{p} \in P \ . \\
(3) & grad\,\lambda_3(\mathbf{p}) & = & (0\,,\,-i\,,\,0)\ . \\
 & \Rightarrow\ Ker\,L_3 & \cong & \mathbb{R}^2 \left(\text{or } \mathbb{C}^2\right) \quad \forall\ \mathbf{p} \in P \ .
\end{array}
$$

$\triangle$

If f_k is a degenerated direction and the HESSEian $D^2 f_k(\mathbf{0},\mathbf{p})$ has the corank $\overline{r}$, then – according to chapter 2.4.1 – $\overline{r}$ critical variables exist which describe the degenerated directions of the coordinates. Without any restriction the critical variables are $\overline{\mathbf{x}} := (x_1, \ldots, x_{\overline{r}})$. Then the HESSEian is

$$
D^2 f_k(\mathbf{0},\mathbf{p}) = \begin{array}{c|c|c|}
 & \overline{r} & n-\overline{r} \\
\hline
\overline{r} & \mathbf{0} & \mathbf{0} \\
\hline
n-\overline{r} & \mathbf{0} & \blacksquare \\
\hline
\end{array} \tag{2.71}
$$

The type of the surface described by the noncritical variables of the function f_k in the neighborhood of the singularity is uniquely determined. Therefore it is sufficient to investigate the behavior of f_k only with respect to the subspace described by the critical coordinates $x_1, \ldots, x_{\overline{r}}$. If $\overline{f}_k$ represents the degenerated direction by neglecting the noncritical variables $x_{r+1}, \ldots, x_n$, then the TAYLOR-expansion of $\overline{f}_k$ is

$$
\overline{f}_k(\overline{\mathbf{x}},\mathbf{p}) = \sum_{\substack{\mu \\ |\mu| \geq 3}} \alpha_{\mu,\mathbf{k}}\, \overline{\mathbf{x}}^{\mu} \ . \tag{2.72}
$$

That means the terms of $\overline{f}_k$ begin with the degree ρ.

In order to define the codimension, which is necessary to classify the singularity, the following set of functions is needed:

$$
m^s_{\mathbf{p}}(\overline{r}) := \left\{ g(\cdot,\mathbf{p}) : \mathbb{R}^{\overline{r}} \to \mathbb{R}, \text{analytical} \;\middle|\; \overline{D}^{|\mu|} g(\mathbf{0},\mathbf{p}) = 0 \quad \forall\ |\,\boldsymbol{\mu}\,| < s \right\} \tag{2.73}
$$

$$\text{with} \quad (1) \quad s \in \mathbb{N} \ ,$$

$$(2) \quad \boldsymbol{\mu} := (\mu_1, \ldots, \mu_n) \ , \quad |\, \boldsymbol{\mu} \,| := \sum_{i=1}^{n} \mu_i$$

$$\text{and} \quad (3) \quad \overline{D}^{|\mu|} g := \frac{\partial^{|\mu|} g}{\partial x_1^{\mu_1} \partial x_2^{\mu_2} \ldots \partial x_{\overline{r}}^{\mu_{\overline{r}}}} \ . \tag{2.74}$$

According to equation (2.74) we obtain

$$\overline{f}_k(\cdot, \mathbf{p}) \ \in \ m_{\mathbf{p}}^3(\overline{r}) \ ^6 \tag{2.75}$$

$m_{\mathbf{p}}^s(n)$ contains all functions whose TAYLOR-expansion begins with terms of degree greater than or equal to s. Additionally, the set

$$< D\overline{f}_k(\cdot, \mathbf{p}) > := \left\{ \sum_{i=1}^{\overline{r}} \left(g_i \cdot \frac{\partial \overline{f}_k}{\partial x_i} \right) (\cdot, \mathbf{p}) \middle| \, g_i : \mathbb{R}^{\overline{r}} \to \mathbb{R}, \text{analytic} \right\} \tag{2.76}$$

determined by the gradient of $\overline{f}_k$ at $\overline{\mathbf{x}}$ has to be defined.
For each $k \in \mathbb{N}$ there is a $j \in \mathbb{N}$ so that

$$< D\overline{f}_k(\cdot, \mathbf{p}) > \ \subset \ m_p^j(\overline{r}) \tag{2.77}$$

is true, because each degenerated function $\overline{f}_k(\cdot, \mathbf{p})$ begins with terms of degree greater than or equal to 3.

The polynomial vectorspace $\Pi_{\overline{r}}^n$, which is defined by the set of all polynomials consisting of monomials $\alpha_{\boldsymbol{\mu}} \overline{\mathbf{x}}^{\boldsymbol{\mu}}$ $(1 \leq |\, \mu \,| \leq n)$, $\overline{\mathbf{x}} \in \mathbb{R}^{\overline{r}}$, can be identified with $\mathbb{R}^N$, where $N(n) \in \mathbb{N}$ is an appropriate number. By that identification each polynomial coefficient represents a direction of the real space $\mathbb{R}^N$. Hence, each point of $\mathbb{R}^N$ represents exactly one polynomial.

The set of all polynomials $g \in \Pi_{\overline{r}}^n$ for which the derivatives $\overline{D}^{|\mu|} g(\mathbf{0})$ disappear, where $\boldsymbol{\mu}$ is an element of the set

$$I := \left\{ \boldsymbol{\mu} := (\mu_1, \ldots, \mu_{\overline{r}}) \in \mathbb{N}^{\overline{r}} \middle| \, |\, \boldsymbol{\mu} \,| < n \right\} \tag{2.78}$$

of indices, defines a geometrical object of codimension $|\, I \,|$ in the space $\mathbb{R}^N$, namely the set of zeros:

[6] Strictly spoken, the elements of $m_{\mathbf{p}}^s(n)$ are not functions $g(\cdot, \mathbf{p})$ but germs of functions $[g(\cdot, \mathbf{p})]$ in the neighborhood of the singularity, that means an equivalence class, whose relation on $m_{\mathbf{p}}^s(n)$ is defined by

$$g_1 \sim g_2 :\Leftrightarrow \exists \text{ in the neighborhood of } U(0) \text{ so that } g_1(\mathbf{x}) = g_2(\mathbf{x}) \ \forall \ \mathbf{x} \in U(\mathbf{0})$$

However, from a practical point of view the distinction is not necessary and it may be ignored.

$$N_I := \left\{ \boldsymbol{\alpha} \in \mathbb{R}^N \middle| \alpha_\mu = 0 \quad \forall\ \boldsymbol{\mu} \in I \right\} \ . \tag{2.79}$$

The vector $\boldsymbol{\alpha}$ contains all coefficients of a polynomial g. The codimension does not depend on the dimension N of $\mathbb{R}^N$. After these preliminary words, the codimension of the function $\overline{f}_k(\cdot, \mathbf{p})$ can be defined as follows:

Definition (2.1):

The codimension of the function $\overline{f}_k(\cdot, \mathbf{p})$, abbreviated *codim* $\left(\overline{f}_k(\cdot, \mathbf{p})\right)$, is the codimension of the polynomial space

$$X := < D\overline{f}_k(\cdot, \mathbf{p}) >^{7} \tag{2.80}$$

with respect to the space $\mathbb{R}^N$, where N is sufficiently big. The monomials $\overline{\mathbf{x}}^\mu$, whose coefficients $\alpha_{\mu,k}$ do not occur, define a frame

$$B_k := \{ \overline{\mathbf{x}}^\mu \mid \alpha_{\mu,k} = 0 \ ; \quad | \boldsymbol{\mu} | \geq 1 \} \tag{2.81}$$

of the complementary polynomial space $\mathbb{R}^N - N_I$.

○

▽

Example 2.8:

$$\overline{r} = 1 \ , \ x := x_1 \ ; \ \overline{f}_k(x, \mathbf{p}) = x^k \ ; \quad k \in \mathbb{N} \ .$$

$$g(x) = a_0 + a_1 x + a_2 x^2 + \ldots \ .$$

$$\begin{aligned} \Rightarrow \left(\sum_{i=1}^{\overline{r}} g_i \cdot \frac{\partial \overline{f}_k}{\partial x_i} \right) (x, \mathbf{p}) &= g(x) \frac{\partial \overline{f}_k}{\partial x} (x, \mathbf{p}) \ , \\ &= (a_0 + a_1 x + \ldots) \ k \cdot x^{k-1} \ , \\ &= (k \, a_0) \, x^{k-1} + (k \, a_1) \, x^k + \ldots \ , \\ &= \alpha_{k-1} \, x^{k-1} + \alpha_k \, x^k + \ldots \ . \end{aligned}$$

[7] Since each polynomial $\sum_\mu \boldsymbol{\alpha}_\mu \mathbf{x}^\mu = \alpha_0 + \mathbf{a}_1^T \mathbf{x} + \mathbf{x}^T \mathbf{A}_2 \mathbf{x} + \ldots$ ($\alpha_0 \in \mathbb{R}$; $\mathbf{a}_1 \in \mathbb{R}^n$; $\mathbf{A}_2 \in \mathbb{R}^{n,n}, \ldots$) may be transformed into $\mathbf{z}^T \mathbf{B}_2 \mathbf{z} \ldots$ by $\mathbf{x} = \mathbf{T}\mathbf{z} + \mathbf{b}$, it is necessary that the definition of the codimension does not depend on that property. Therefore a "right shift" of the polynomials $g \in < D\overline{f}(\cdot, p) >$ by multipling them with a polynomial $h \in m_{\mathbf{p}}(\overline{r})$ must be made. I.e. to define the codimension the residue class ring

$$m_{\mathbf{p}}(\overline{r}) / < Df_k(\cdot, \mathbf{p}) >$$

is used, which is a polynomial space expressing the "right shift" mathematically. However, for practical use the definition mentioned above is more convenient.

$$\Rightarrow \quad \alpha_1 = \ldots = \alpha_{k-2} = 0 \qquad \Rightarrow \quad codim\left(\overline{f}_k(\cdot, p)\right) = k - 2 \quad ,$$

and $B_k = \left\{x\,,\ x^2\,,\ x^3\,, \ldots,\ x^{k-2}\right\}$.

△

▽

<u>Example 2.9:</u>

$$\overline{r} = 2\,;\ \ x := x_1\,;\ \ y := x_2\ ,\quad \overline{f}_k(x, y, \mathbf{p}) := x^4 + y^4\ .$$

$$g_1(x,y) \quad = \quad a_0 + a_1 x + a_2 y + s_3 x^2 + a_4 xy + a_5 y^2 + \ldots\ ,$$

$$g_2(x,y) \quad = \quad b_0 + b_1 x + b_2 y + b_3 x^2 + b_4 xy + b_5 y^2 + \ldots\ .$$

$$\Rightarrow \quad \left(\sum_{i=1}^{\overline{r}} g_i \cdot \frac{\partial \overline{f}_k}{\partial x_i}\right)(x, y, \mathbf{p}) \quad = \quad 4x^3\left(a_0 + a_1 x + a_2 y + \ldots\right) + \\ +4y^3\left(b_0 + b_1 x + b_2 y + \ldots\right)\ ,$$

$$= \quad \alpha_6 x^3 + \alpha_9 y^3 + \alpha_{10} x^4 + \alpha_{11} x^3 y + \\ +\alpha_{13} x y^3 + \ldots\ .$$

The coefficients of terms with degree 4 and higher do all exist. The only zero-coefficients are

$$\alpha_1 = \alpha_2 = \alpha_3 = \alpha_4 = \alpha_5 = \alpha_7 = \alpha_8 = \alpha_{12} = 0$$

with

$$B_k = \left\{x\,,\ y\,,\ x^2\,,\ xy\,,\ y^2\,,\ x^2 y\,,\ xy^2\,,\ x^2 y^2\right\}$$

$$\Rightarrow \quad codim\left(f_k(\cdot, \mathbf{p})\right) = 8\ .$$

△

▽

<u>Example 2.10:</u>

$$\overline{r} = 2\,;\quad x := x_1\,;\quad y := x_2\,,\ \overline{f}_k(x, y, p) := x^2 y\ .$$

$$\Rightarrow \qquad codim\ (f_k^*(\cdot, \mathbf{p})) = \infty$$

with

$$B_k = \left\{y,\ y^2,\ y^3, \ldots\right\} \ .$$

In that case all coefficients α_μ of $y^k \forall k \in \mathbb{N}$ disappear, i.e. also the dimension N of $\mathbb{R}^N$ has to be infinite.

△

▽

Example 2.11:

For the direction f_1 of the oscillator described in example 2.7 we have $\overline{f}_1(\overline{\mathbf{x}}, \mathbf{p}) = (\sum t_{1j}(\mathbf{p})x_j)^3$ and $\overline{r} = 4$ (because of $D^2 f_1(\mathbf{0}, \mathbf{p}) = \mathbf{0}$).
If all terms $t_{1p}(\mathbf{p})$ do exist then:

$$codim\ (f_1(\cdot, \mathbf{p})) = 10 \ ,$$

$$B_k = \{x_1\,,\ x_2\,,\ x_3\,,\ x_4\,,\ x_1x_2\,,\ x_1x_3\,,\ x_2x_3\,,\ x_2x_4\,,\ x_3x_4\} \ .$$

△

In general (cf. [ARNOLD 1983], [LU 1976]) the determination of the "codimension" and "corank" is the most essential task to classify the singularities. These definitions (codimension and corank) determine the type of the surface in the neighborhood of the singularity which is described by the degenerated direction f_k. A further aspect has be taken into account: The investigation of the singularities with codimension ≤ 4 and corank ≤ 2 shows that different types inside this class exist. These types have to be determined separately. For instance, each degenerated direction f_k with codimension 3 and corank 2 is either diffeomorph to the "elliptic umbilic" ($f_k \cong x_1^3 - x_1x_2^2$) or to the "hyperbolic umbilic" ($f_k \cong x_1^3 + x_2^2$). The determination of diffeomorph types of degenerated singularities with codimension ≥ 10 and corank > 3, such as in example 2.11, requires a lot of energy and has not been done (cf. [ARNOLD 1985]) yet. In many cases it is possible to get rid of the degeneration of the singularity by variation of the parameter $\mathbf{p}$. This modification is called "unfolding of a singularity". The necessary parameters are called "unfolding parameters".

The unfolding is called "universal", if the number of unfolding parameters is equal to the codimension of the singularity. In other words, the codimension is the smallest number of parameters which are necessary to unfold the singularity completely or universally.

If the degenerated function f_k depends on the parameters $\mathbf{p}$, the family of functions $f_k(\mathbf{x}, \mathbf{p})$ will be modified by variation of $\mathbf{p}$. This modification is called "(uni-)versal deformation" of the family. For instance, in example 2.3 the singularity x^3 (with *codim* 1) has been universally deformed by introducing a new parameter ε. I.e. the unfolding of a degenerated singularity is feasible, if additional parameters are introduced.

However, if the investigation of singularities is based only on the family $f_k(\mathbf{x}, \mathbf{p})$, denoted by the parameter $\mathbf{p}$, then only those singularities are important whose codimension $codim(f_k)$ is less than or equal to the dimension of the linear space $Im(L_k)$ (cf. equation (2.74). That means the codimension must be less than or equal to the number of parameters p_i which influence the singularity. Otherwise, the singularity of the family $f_k(\mathbf{x}, \mathbf{p})$ can be removed by an arbitrary small deformation.

▽

Example 2.12:

The family $f_k(x, \mathbf{p}) = p_1 x^3 + p_2 x^4 + \ldots$ has a degenerated singularity at $x = 0$ with $codim(f_k) = 1$. It is not possible to unfold this singularity by variation of $p_1, p_2, \ldots$. However, if a new parameter ε is introduced such that

$$\hat{f}_k(x, \mathbf{p}, \varepsilon) = \varepsilon x + p_1 x^3 + p_2 x^4 + \ldots \ ,$$

then a versal deformation removes the degeneration for each $\varepsilon > 0$.

△

▽

Example 2.13:

Consider the family

$$f_k(x, \mathbf{p}) = p_0 x + p_1 x^3 + p_2 x^4 + \ldots \ .$$

The singularity $x = 0$ is degenerated for $p_0 = 0$ (codimension 1). However, the singularity stays:

$$\hat{f}_k(x, \mathbf{p}, \varepsilon) = (p_0 + \varepsilon)\, x + p_1 x^3 + \ldots \ ,$$

since for each $\varepsilon > 0$, $x = 0$ is degenerated again if we chose $p_0 = -\varepsilon$.

△

Therefore, degenerated singularities of a family of functions $f_k(\mathbf{x}, \mathbf{p})$ are only of practical interest if

$$codim\,(f_k(\cdot, \mathbf{p})) \leq dim\,(L_k(T_\mathbf{p}P)) \tag{2.82}$$

is true.

2.4.3 Structural stability

If a dynamical system $\dot{\mathbf{x}} = \mathbf{f}(\mathbf{x}, \mathbf{p})$ is investigated numerically, we have to make sure that small disturbances (versal deformations) of the system lead to approximately the same results. The same is the case for an experiment where the parameters are never exactly determined. Therefore it is reasonable to call a dynamical system $\dot{\mathbf{x}} = f(\mathbf{x}, \mathbf{p})$ or a family of functions $\mathbf{f}(\mathbf{x}, \mathbf{p})$ "structurally stable" if a small versal deformation leads to "similar solutions" or solutions with "same properties".

For example, if $\mathbf{x} = \mathbf{0}$ is a singular point of the vector field $\mathbf{f}$ $(\mathbf{f}(\mathbf{0}, \mathbf{p}) = \mathbf{0})$ and

$$\lambda_k = \alpha_k + i\beta_k \qquad \text{with} \quad \alpha_k,\ \beta_k \neq 0 \tag{2.83}$$

are the eigenvalues of $D\mathbf{f}(\mathbf{0}, \mathbf{p})$, then an open neighborhood $U \subset \mathbb{C}^{2f}$ of $\boldsymbol{\lambda} := (\lambda_1, \ldots, \lambda_{2f})$ exists so that each $\hat{\boldsymbol{\lambda}} \in U$ is a pure complex number, i.e.

$$\hat{\lambda}_k = \hat{\alpha}_k + i\hat{\beta}_k \qquad \text{with} \quad \hat{\alpha}_k,\ \hat{\beta}_k \neq 0\ . \tag{2.84}$$

In that case, the solutions of the linear system representing the linear part of the non-linear system around $\mathbf{x} = \mathbf{0}$, as well as the solutions of the non-linear system (theorem of GROBMANN and HARTMAN) keep their structure after a versal deformation. With respect to that singularity the family $\mathbf{f}(\mathbf{x}, \mathbf{p})$ is structurally stable.

In the same way, a family may be structurally stable with respect to a degenerated singularity (cf. example 2.13).

In case of $\lambda_k = 0$ or $\lambda_k = i\beta_k$ each arbitrary small disturbance leads to a pure complex eigenvalue λ_k with $Re\lambda_k \neq 0$ and therefore to qualitatively different solutions in the neighborhood of $\mathbf{x} = \mathbf{0}$. According to the dimension given above the family $\mathbf{f}(\mathbf{x}, \mathbf{p})$ is not structurally stable. The same statement is true for example 2.12.

That means a dynamical system $\dot{\mathbf{x}} = \mathbf{f}(\mathbf{x}, \mathbf{p})$ is structurally stable in the neighborhood of a singularity, iff equation (2.82) is fulfilled for all $k \in \{1, \ldots, 2f\}$, no matter whether the singularity is degenerated or not.

2.5 Periodic solutions of HAMILTONian systems

... welche auf das n-Körper-Problem anwendbar sind und darüber hinaus sogar auf viel allgemeinere Fragen der Mechanik. Dabei handelt es sich um die Bestimmung periodischer Lösungen ...

Carl Ludwig Siegel [8]

The investigation of periodic solutions of frictionless nonexcited dynamical systems, which we want to discuss in this section, seems – especially from a practical point of view – not very important, since they are only exceptions or a "zero set" among all solutions of non-linear systems. However, the necessity of studying periodic solutions will become more evident, if, for instance, example 2.2 is considered again. All solutions of the non-linear system (double pendulum) were determined by the solutions of the linear part obtained by linearization of the system around the stable equilibrium point. Any solution $\mathbf{x}$ of the linear system was a linear combination of the eigenvectors $\mathbf{v}_k$ multiplied with the basic functions $e^{(\delta_k+i\omega_k)t}$:

$$\mathbf{x}(t) = \sum_{k=-2}^{2} c_k \mathbf{v}_k e^{(\delta_k+i\omega_k)t} \tag{2.85}$$

(with $\omega_{-k} = -\omega_k$ and $\mathbf{v}_{-k} = -\overline{\mathbf{v}}_k$).

If the influence of the friction (described by δ_k) disappears, the linear system will become HAMILTONian and each solution consists of a linear combination of periodic solutions. However, for $\delta_k \to 0$ the system will be structurally unstable and resonant in the neighborhood of the equilibrium point. According to the theory of normal forms (cf. chapter 2.3) the resonance leads to the loss of the diffeomorph equivalence between the non-linear system and the linearized system. Nevertheless it may be assumed – particularly according to the proofs of [WEINSTEIN 1973] and [MOSER 1976] – that there is an immediate connection between the periodic solutions of the linearized system and the periodic solutions of the non-linear system in the neighborhood of the singularity (equilibrium point).

From a practical point of view it is important to know that in numerical computations of large non-linear systems an (randing errors, ill conditioned etc.) eigenvalue $\lambda_k = \delta_k + i\omega_k$ with $0 < \delta_k < \varepsilon_{mach}$ may always be interpreted as a pure imaginary eigenvalue which causes resonancy. Therefore an almost periodic solution will turn out to be a periodic one, and an almost HAMILTONian system will be practically HAMILTONian.

[8] From [SIEGEL 1956] p.69

In most cases of application the friction is relatively small. Therefore, it is more convenient to investigate the neighboring HAMILTONian system, which has more mathematical structure than the real system and transfer the results to the real system.

2.5.1 The boundary value problem for periodic solutions of HAMILTONian systems

The vector field $\mathbf{f} : TM \times P \to TM$ (cf. equation (2.27)) of HAMILTONian systems, which describes the dynamical behavior, has always a first integral, namely the HAMILTONian function $H \in C^\infty(TM \times P, \mathbb{R})$. H can always be expressed explicitly.

However, the fact that the HAMILTONian function H is constant along an arbitrary trajectory is not an algebraic constraint such as equation (2.3). It is rather an algebraic restriction which defines a state space of codimension 1. That is

$$H(\mathbf{x}, \mathbf{p}) = h = const. \tag{2.86}$$

defines a $(2f-1)$–dimensional differentiable manifold

$$H^{-1}(h) := \left\{ \mathbf{x} \in \mathbb{R}^{2f} \middle| H(\mathbf{x}, \mathbf{p}) - h = 0 \right\} , \tag{2.87}$$

which includes all trajectories of the dynamical system.

The use of impulse-coordinates

$$\mathbf{z} := \begin{bmatrix} \mathbf{q} \\ \mathbf{M}(\mathbf{q}) \cdot \dot{\mathbf{q}} \end{bmatrix} = \begin{bmatrix} \mathbf{E} & \mathbf{0} \\ \mathbf{0} & \mathbf{M}(\mathbf{q}) \end{bmatrix} \cdot \mathbf{x} \tag{2.88}$$

($\mathbf{M}(\mathbf{q}) \in \mathbb{R}^{f,f}$ massmatrix, $\mathbf{q} \in \mathbb{R}^f$ generalized coordinates) transforms the HAMILTONian system into its standard form

$$\dot{\mathbf{z}} = \mathbf{J} \cdot DH(\mathbf{z}, \mathbf{p}) \tag{2.89}$$

with the symplectic matrix

$$\mathbf{J} := \begin{bmatrix} \mathbf{0} & \mathbf{E} \\ -\mathbf{E} & \mathbf{0} \end{bmatrix} \tag{2.90}$$

and the gradient of H

$$DH(\mathbf{z}, \mathbf{p}) := \frac{\partial H}{\partial \mathbf{z}}(\mathbf{z}, \mathbf{p}) . \tag{2.91}$$

With respect to numerical investigations equation (2.27) is more convenient than equation (2.89). But equation (2.89) has more advantages in terms of theoretical investigations.

The objective of the following section concerns the numerical computation of periodic solutions and the investigation of the stability and bifurcation behavior of these solutions.

A survey of the existence of periodic solutions of non-linear systems can be found in [RABINOWITZ 1982], [RABINOWITZ 1978] or [DUISTERMAAT 1984]. In this chapter we assume the existence of periodic solutions if not explicitly mentioned.

The numerical computation of periodic solutions leads to a two point boundary value problem (BVP). The BVP is given by the ODE according to equation (2.27) or (2.89) respectively and by the boundary values

$$\mathbf{r}(\mathbf{x}(0),\, \mathbf{x}(T)) \;:=\; \mathbf{x}(T) - \mathbf{x}(0) - 2\mathbf{k}\pi = \mathbf{0} \tag{2.92}$$

or

$$\mathbf{r}(\mathbf{z}(0),\, \mathbf{z}(T)) \;:=\; \mathbf{z}(T) - \mathbf{z}(0) - 2\mathbf{k}\pi = \mathbf{0} \tag{2.93}$$

respectively.

$T \in \mathbb{R}$ is the time of period which is still unknown and part of the computation. $\mathbf{k} = (k_1, \ldots, k_f\ ,\ 0, \ldots, 0) \in \mathbb{N}^{2f}$ is an integer vector which denotes the number of rotations of the coordinates x_j inside one period.

∇

Example 2.14: (double pendulum without friction)

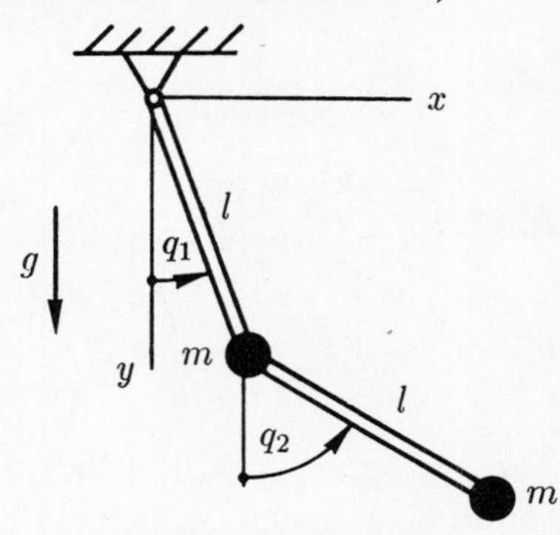

Fig. 2.6: frictionless double pendulum

If the dampers of the double pendulum are removed, the system will become conservative or HAMILTONian.

- The HAMILTONian function is (same masses m and lengths l):

$$H(\mathbf{x},\mathbf{p}) := ml^2\left[x_3^2 + \tfrac{1}{2}x_4^2 + x_3x_4\cos(x_1 - x_2)\right] + $$
$$+ml^2\omega_0^2\left[3 - 2\cos x_1 - \cos x_2\right] \ .$$

$$\mathbf{x} := (q_1\,,\, q_2\,,\, \dot{q}_1\,,\, \dot{q}_2)^T$$

is the state vector and

$$\mathbf{p} := (m\,,\, l\,,\, \omega_0)^T \qquad \left(\omega_0^2 := \frac{g}{l}\right)$$

the parameter vector.

- The equations of motion expressed in coordinates of the state space are

$$\dot{\mathbf{x}} = \mathbf{f}(\mathbf{x},\mathbf{p}) := \begin{bmatrix} x_3 \\ x_4 \\ M^{-1}(\mathbf{x})\,\mathbf{r}(\mathbf{x},\mathbf{p}) \end{bmatrix}$$

with the mass matrix

$$\mathbf{M}(\mathbf{x}) := \begin{bmatrix} 2 & \cos(x_1 - x_2) \\ \cos(x_1 - x_2) & 1 \end{bmatrix}$$

and the right hand side

$$\mathbf{r}(\mathbf{x},\mathbf{p}) := \begin{bmatrix} -x_4^2 & \sin(x_1 - x_2) & -2\omega_0^2 \sin x_1 \\ x_3^2 & \sin(x_1 - x_2) & -\omega_0^2 \sin x_2 \end{bmatrix} \ .$$

- The periodicity condition is

$$\mathbf{x}(T) = \mathbf{x}(0) + 2\mathbf{k}\pi$$

with

$$\mathbf{k} = (k_1\,,\, k_2\,,\, 0\,,\, 0)^T \quad ; \quad k_1\,,\, k_2 \in \mathbb{Z} \ .$$

k_1 and k_2 are the numbers of rotations of pendulum 1 and pendulum 2 necessary to come back to the initial point of the trajectory after one period.

△

In general, the BVP of HAMILTONian systems is described by

$$\boxed{\begin{aligned} \dot{\mathbf{x}} &= \mathbf{f}(\mathbf{x},\mathbf{p}) \ , \\ H(\mathbf{x},\mathbf{p}) &= h \quad (= const.) \ , \\ \mathbf{x}(T) &= \mathbf{x}(0) + 2\mathbf{k}\pi \ . \end{aligned}} \tag{2.94}$$

2.5.2 The problem of the translation invariance of periodic solutions of autonomous systems

If $\mathbf{x} : \mathbb{R} \to \mathbb{R}^{2f}$ is a solution of the BVP equation (2.94), $\mathbf{x} \circ \tau$ is a solution for any translation $\tau(t) := t + \Delta t$, $\Delta t \in \mathbb{R}$ too. The BVP has no unique solution since the time t does not explicitly occur in the boundary condition.

However, for dynamical systems there is a possibility to get rid of this problem by using the mean value theorem for a configuration space coordinate x_j.

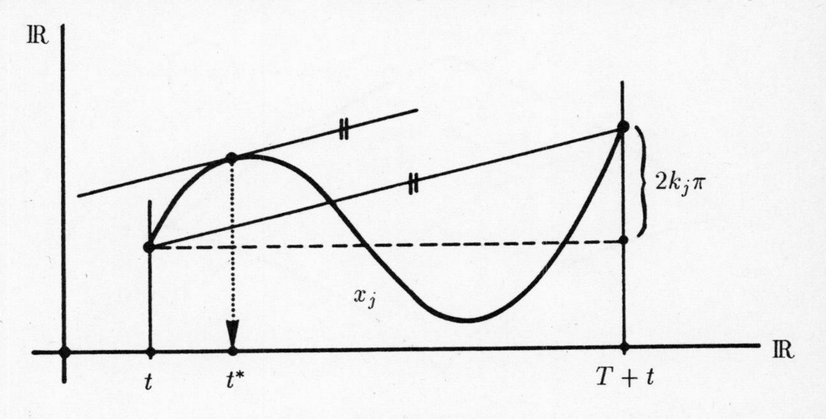

Fig. 2.7: application of the mean value theorem to the coordinate x_j

If x_j is the j-th generalized coordinate, then there is an isolated point t^* inside the interval $[t\,,\,T+t]$ of one period at which the value of the derivative $\dot{x}_j$ is equal to the increasing value $x_j(T) - x_j(0)$ of the function x_j with respect to T, that is

$$\exists\ t^* \in [t\,,\,T+t] \ \text{ such that } \ \dot{x}_j(t^*) = \frac{x_j(T) - x_j(0)}{T} \ . \tag{2.95}$$

Equation (2.95) can be simplified if the periodicity condition $x_j(T) = x_j(0) + 2k_j\pi$ and the properties $\mathbf{x} = \left(\mathbf{q}^T, \dot{\mathbf{q}}^T\right)$ as well as $\dot{x}_j = x_{j+f}$ are taken into account:

$$x_{j+f}(t^*) = \frac{2k_j\pi}{T} \quad . \tag{2.96}$$

Because of the invariance of translation we choose $t^* = 0$. That means t^* is the initial time and

$$x_{j+f}(0) = \frac{2k_j\pi}{T} \quad . \tag{2.97}$$

Equations (2.97) and (2.94) provide a unique solution in case the solution exists and is locally separated in the state space.

In HAMILTONian systems the periodic solutions are not separated if there is a periodic solution for each value h of the HAMILTONian function H. In that case, it is necessary to express the system in surface oriented coordinates of the manifold $H^{-1}(h)$. On this manifold the periodic solutions are usually separated.

Assume that $\eta_1, \ldots, \eta_{2f-1}$ are surface orientated local coordinates on $H^{-1}(h)$ and

$$S_h \; : \; H^{-1}(h) \;\; \rightarrow \;\; \mathbb{R}^{2f} \;\; , \quad \boldsymbol{\eta} \mapsto \mathbf{x} \tag{2.98}$$

defines an embedding of $H^{-1}(h)$ into $\mathbb{R}^{2f}$, then,

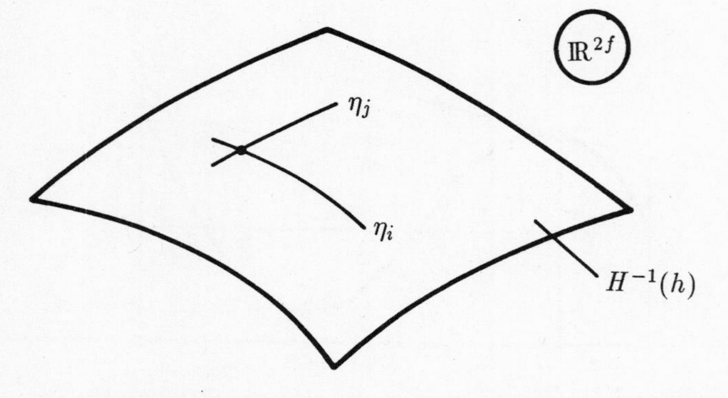

Fig. 2.8: surface orientated coordinates on the energy surface $H^{-1}(h)$

employing the chain rule equation (2.94) will be transformed into

$$DS_h(\boldsymbol{\eta}) \cdot \dot{\boldsymbol{\eta}} = \mathbf{f}\,(S_h(\boldsymbol{\eta})\,,\,\mathbf{p}) \quad . \tag{2.99}$$

The rank of $DS_h(\boldsymbol{\eta}) \in \mathbb{R}^{2f,2f-1}$ is $2f-1$ for each $\boldsymbol{\eta}$. Therefore, one of the $2f$ ODE's in equation (2.99) is redundant. The QR-decomposition of the matrix $DS_h(\boldsymbol{\eta})$ is one possibility to avoid this problem. Namely, if $DS_h(\boldsymbol{\eta}) = \mathbf{Q} \cdot \mathbf{R}$, then we have:

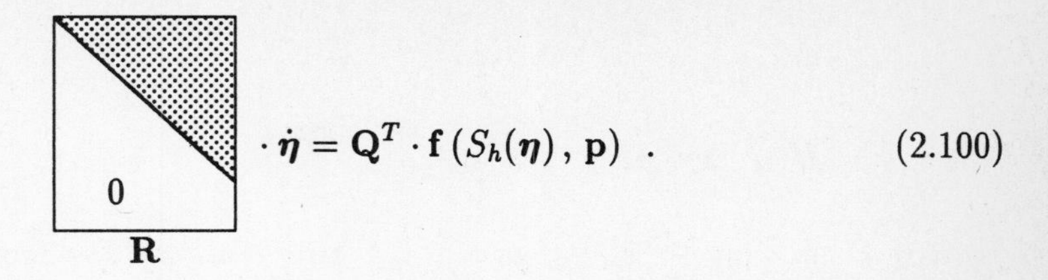

$$\cdot \dot{\boldsymbol{\eta}} = \mathbf{Q}^T \cdot \mathbf{f}\left(S_h(\boldsymbol{\eta}), \mathbf{p}\right) \quad . \tag{2.100}$$

The last ODE in equation (2.100) can be neglected. If we employ the MOORE–PENROSE–inverse matrix $[DS_h(\boldsymbol{\eta})]^+$ of $DS_h(\boldsymbol{\eta})$ we get the reduced system:

$$\begin{aligned} \dot{\boldsymbol{\eta}} &= [DS_h(\boldsymbol{\eta})]^+ \mathbf{f}\ (S_h(\boldsymbol{\eta}), \mathbf{p}) \\ &=: \hat{\mathbf{f}}(\boldsymbol{\eta}, \hat{\mathbf{p}}) \ . \end{aligned} \tag{2.101}$$

The value h of the HAMILTONian function H is a parameter of the system. The parameter vector $\hat{\mathbf{p}}$ contains $\mathbf{p}$ and h as well.

The corresponding boundary conditions are:

$$\mathbf{r}(\boldsymbol{\eta}(0), \boldsymbol{\eta}(T)) := S_h(\boldsymbol{\eta}(T)) - S_h(\boldsymbol{\eta}(0)) - 2\mathbf{k}\pi = \mathbf{0} \ . \tag{2.102}$$

According to the implicit function theorem equation (2.102) has a unique solution $\boldsymbol{\eta}(0)$ iff the JACOBI'an $\frac{\partial \mathbf{r}}{\partial \boldsymbol{\eta}(0)}(\boldsymbol{\eta}(0), \boldsymbol{\eta}(T))$ has the rank $2f-1$. This is the case since we have

$$\frac{\partial \mathbf{r}}{\partial \boldsymbol{\eta}(0)}(\boldsymbol{\eta}(0), \boldsymbol{\eta}(T)) = -DS_h(\boldsymbol{\eta}(0)) \ . \tag{2.103}$$

Therefore, equation (2.102) represents exactly $2f-1$ independent boundary conditions. The boundary condition (2.97) or

$$S_{h,j+f}(\eta(0)) = \frac{2k_j\pi}{T} \tag{2.104}$$

respectively, which is independent from equation (2.103) ($S_{h,j+f}$ is the $(j+f)$-th component of S_h), can be used to determine the time T of period (cf. chapter 2.5.3).

∇

Example 2.15:

If we choose

$$\eta_i := x_i \qquad (i = 1, 2, 3)$$

in example 2.14, then the coordinate x_4 or transformation S_h respectively can be computed from the HAMILTONian function:

$$
\begin{aligned}
h &= H(\mathbf{x}, \mathbf{p}) \;, \\
&= H(\boldsymbol{\eta}, x_4, \mathbf{p}) \;, \\
&= ml^2 \left[\eta_3^2 + \tfrac{1}{2}x_4^2 + \eta_3 x_4 \cos(\eta_1 - \eta_2)\right] + ml^2\omega_0^2 \left[3 - 2\cos\eta_1 - \cos\eta_2\right] \;.
\end{aligned}
$$

This equation is quadratic with respect to x_4 and therefore easy to solve. The method is applicable to any HAMILTONian system since H is always quadratic and positive definite with respect to the coordinates of velocity.

The ODE's are

$$
\begin{aligned}
\dot{\eta}_1 &= \eta_3 = \hat{f}_1(\boldsymbol{\eta}, \hat{\mathbf{p}}) \;, \\
\dot{\eta}_2 &= x_4(\boldsymbol{\eta}) = \hat{f}_2(\boldsymbol{\eta}, \hat{\mathbf{p}}) \;, \\
\dot{\eta}_3 &= \tfrac{1}{2-\cos^2(\eta_1-\eta_2)} \left[-x_4(\boldsymbol{\eta})^2 \sin(\eta_1 - \eta_2) - 2\omega_0^2 \sin\eta_1 - \right. \\
&\qquad \left. (\eta_3^2 \sin(\eta_1 - \eta_2) - \omega_0^2 \sin\eta_2) \cos(\eta_1 - \eta_2)\right] \\
&=: \hat{f}_3(\boldsymbol{\eta}, \hat{\mathbf{p}}) \;.
\end{aligned}
$$

Equation (2.97) is true for each admissable value of i. The boundary condition for $i = 1$ is

$$
\begin{aligned}
\boldsymbol{\eta}(T) &= \boldsymbol{\eta}(0) - 2\pi \begin{bmatrix} k_1 \\ k_2 \\ 0 \end{bmatrix} \;, \\
\eta_3(0) &= \tfrac{2k_1\pi}{T} \;.
\end{aligned}
$$

This formulation ensures the invariance of translation as well as the isolation of the periodic solutions with respect to the energy value h. Furthermore, the formulation can be applied to any HAMILTONian system.

Other approaches such as modified GAUSS–NEWTON techniques – proposed by [DEUFLHARD 1984] – need not to be employed for dynamical systems.

$\triangle$

2.5.3 Formulation of the BVP with fixed endpoint

As mentioned above, the time of period is unknown. Therefore it is necessary to transform the BVP (equations (2.101) to (2.104)) into a BVP with fixed endpoint.

One convenient way to obtain the desired transformation is to define the dimensionless time $\tau := \frac{t}{T}$ and a new variable η_{2f} by:

$$\eta_{2f} := T \ . \tag{2.105}$$

(cf. [STOER, BULIRSCH 1979])

This leads to

$$\begin{aligned} \eta_i' &= \dot{\eta}_i \cdot \tfrac{dt}{d\tau} \ , \qquad \left({}' := \tfrac{d}{d\tau}\right) \\ &= \dot{\eta}_i \cdot \eta_{2f} \end{aligned}$$

as well as

$$\eta_{2f}' = 0 \ . \tag{2.106}$$

Because of $\eta_{2f}(0) = \eta_{2f}(T) = \eta_{2f}(t) = const.$ condition (2.104) for the free endpoint is expressed by

$$S_{h,j+f}(\boldsymbol{\eta}(0)) \cdot \eta_{2f}(0) = 2k_j\pi \ . \tag{2.107}$$

The new time of period has the value 1. Furthermore, if we introduce the new variables $Y_i(\tau) := \eta_i(t)$ a standard BVP results which can be solved by commonly available library routines.

∇

Example 2.16:

If example 2.15 is modified, as mentioned above, the following standard BVP will result:

ODE:	$Y_1' = Y_3 \cdot Y_4$ $Y_2' = x_4 \cdot Y_4$ $Y_3' = \frac{Y_4}{2-\cos(Y_1-Y_2)}[-x_4^2 \sin(Y_1 - Y_2) - 2\omega_0^2 \sin Y_1 -$ $- (Y_3^2 \sin(Y_1 - Y_2) - \omega_0^2 \sin Y_2)\cos(Y_1 - Y_2)]$ $Y_4' = 0$	
GC:	$h = ml^2 \left[Y_3^2 + \frac{1}{2}x_4^2 + Y_3 x_4 \cos(Y_1 - Y_2)\right] +$ $+ ml^2\omega_0^2 [3 - 2\cos Y_1 - \cos Y_2]$	
BV:	$Y_i(1) = Y_i(0) + 2k_i\pi \qquad (i = 1, 2, 3)$ $Y_3(0)Y_4(0) = 2k_1\pi$	

△

2.5.4 Numerical computation of periodic solutions of the double pendulum

"pulchritudo splendor veritatis" [9]

In the following the double pendulum without friction demonstrated in the examples 2.14 – 2.16 will be used for some investigations in which the parameter **p** is varied. The results will be used in chapter 2.8.4 (stability and bifurcation of HAMILTONian systems) in order to give an explanation for the irregular behavior inside certain energy ranges.

Masses and lengths of the pendulums are assumed to be equal ($m_1 = m_2 = m$). The energy values h and the numbers k_j of rotations will be varied.

According to the dataset given in appendix 1 the double pendulum has the following points of equilibrium or singularities respectively:

[9] aesthetics is the splendor of truth

$$\begin{aligned} \mathbf{q}^{(1)} &= (0\,,0) && \text{for} \quad h = h_1 = 0\ , \\ \mathbf{q}^{(2)} &= (0\,,\pi) && \text{for} \quad h = h_2 = 2\ , \\ \mathbf{q}^{(3)} &= (\pi\,,0) && \text{for} \quad h = h_3 = 4 \\ \text{and}\quad \mathbf{q}^{(4)} &= (\pi\,,\pi) && \text{for} \quad h = h_4 = 6\ . \end{aligned}$$

$\mathbf{q}^{(1)}$ is stable, but not asymptotically stable, the other points are unstable. In HAMILTONian systems the bifurcation parameter is the energy value h or the time T of period. If the fixed point is stable, a bifurcation from the trivial periodic solution $\mathbf{q} = \mathbf{0}$ and $T = 0$ to nontrivial periodic solutions $\mathbf{q} \neq \mathbf{0}$ and $0 < T < \infty$ will occur (chapter 2.8). If the fixed point is unstable, a homoclinic orbit $\mathbf{q} \neq \mathbf{0}$ and $T = \infty$ bifurcates (formally) to a nontrivial periodic solution $\mathbf{q} \neq \mathbf{0}$ and $0 < T < \infty$. Homoclinic orbits cannot be numerically computed. Therefore it is convenient to start with periodic solutions of the linearized system around the stable fixed point to compute periodic solutions of the non-linear system. For the numerical computation of the periodic solutions the multiple shooting method (cf. [BULIRSCH 1971], [DIEKHOFF 1979]) will be applied.

The energy parameter h is used as homotopy parameter, that means: if a periodic solution which belongs to the energy value h is found, it will be used as an initial function to find the periodic solution for the energy value $h + \Delta h$.

<u>1st case:</u> $k_1 = 0$; $k_2 = 0$

Only periodic librations of pendulum 1 as well as pendulum 2 are considered. The linearized system around $\mathbf{q}^{(1)}$ has two solutions or eigenmodes respectively.

(a) <u>initial function:</u>

$$\begin{aligned} Y_1^{(0)}(\tau) &= A\cos(2\pi\tau)\ , & Y_3^{(0)}(\tau) &= -A\omega\sin(2\pi\tau)\ , \\ Y_2^{(0)}(\tau) &= \sqrt{2}\,A\cdot\cos(2\pi\tau)\ , & Y_4^{(0)}(\tau) &= \tfrac{2\pi}{\omega} \end{aligned}$$

with

$$A := 0.1745\,rad \quad (\hat{=}\,10^\circ)\ , \qquad \omega := 0.765\,\omega_0\ .$$

To generate the homotopy mentioned above, the energy values will be increased in discrete steps $\Delta h = 0.1$.

The figures 2.9 and 2.10 show the projections of the periodic solutions into the phase planes $(q_1, \dot{q}_1)$ and $(q_2, \dot{q}_2)$ for different steps of energy.

As expected, the solution $\mathbf{q}(t)$ around the fix point is like the solution of the linearized system around the fix point. The solution $\mathbf{q}(t)$ will be deformed with increasing energy. The solutions can be generated up to energy values $h < h^{(4)}$.

For $h \to h^{(4)}$ the periodic solution will be deformed into a homoclinic orbit $(T \to \infty)$ or into the fix point $\mathbf{q}^{(4)}$ respectively. There is no solution of this type for higher values of energy.

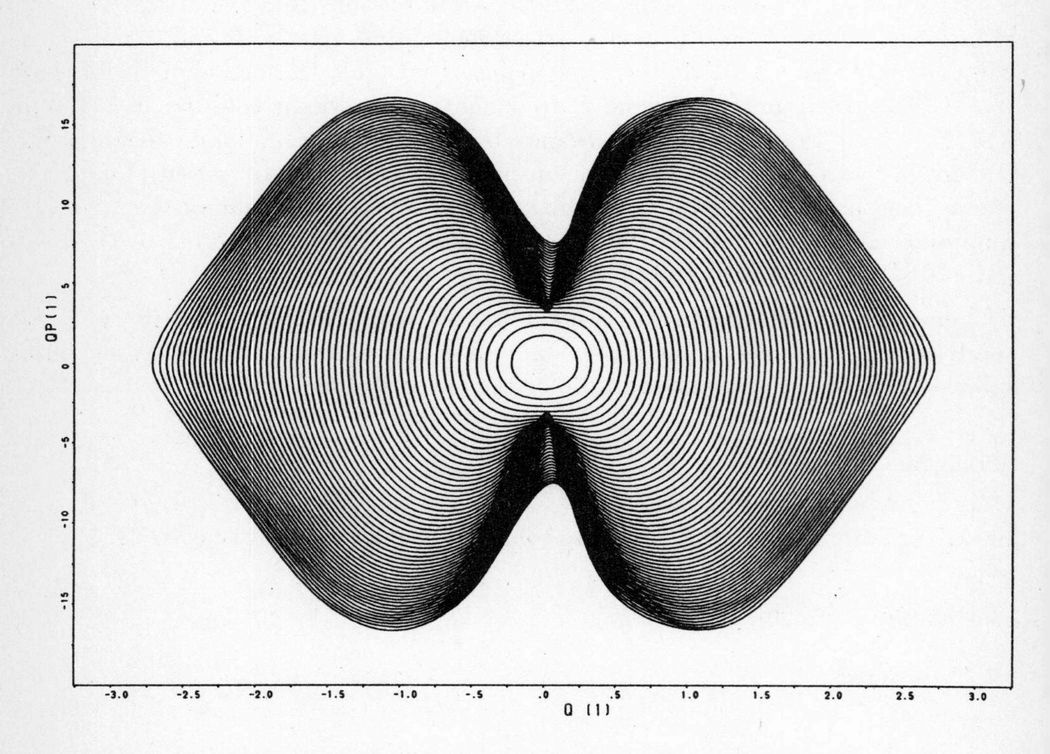

Fig. 2.9: phase curves $(q_1, \dot{q}_1)$ of the first branch of periodic solutions $\left(0 < h < h^{(4)}\right)$

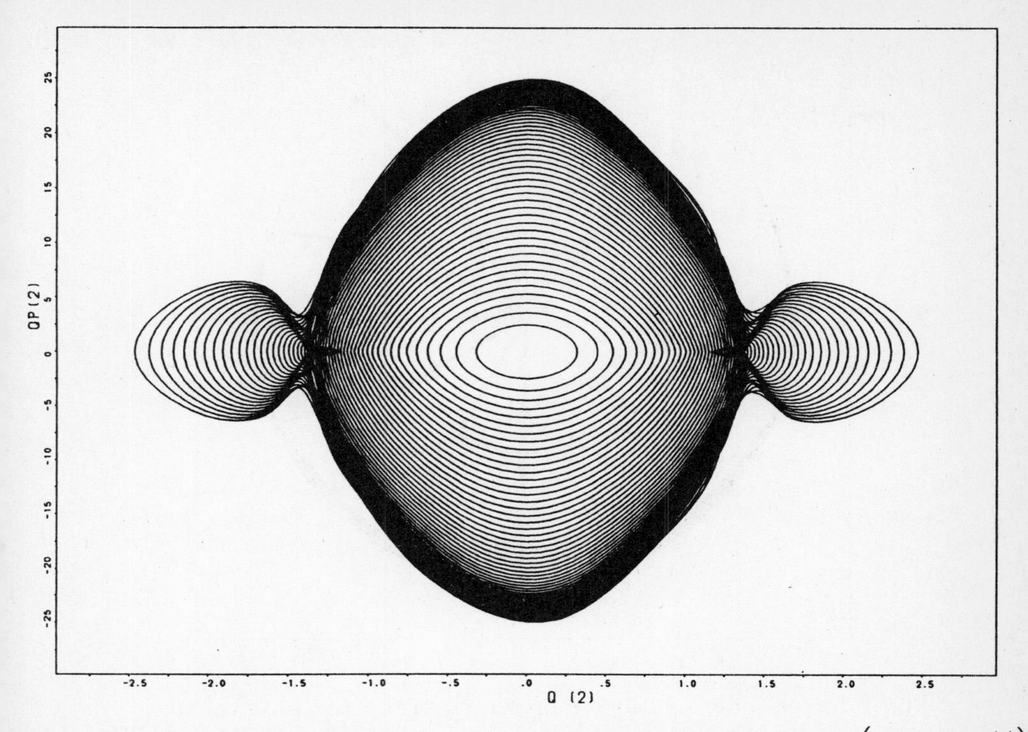

Fig. 2.10: phase curves $(q_2, \dot{q}_2)$ of the first branch periodic solutions $\left(0 < h < h^{(4)}\right)$

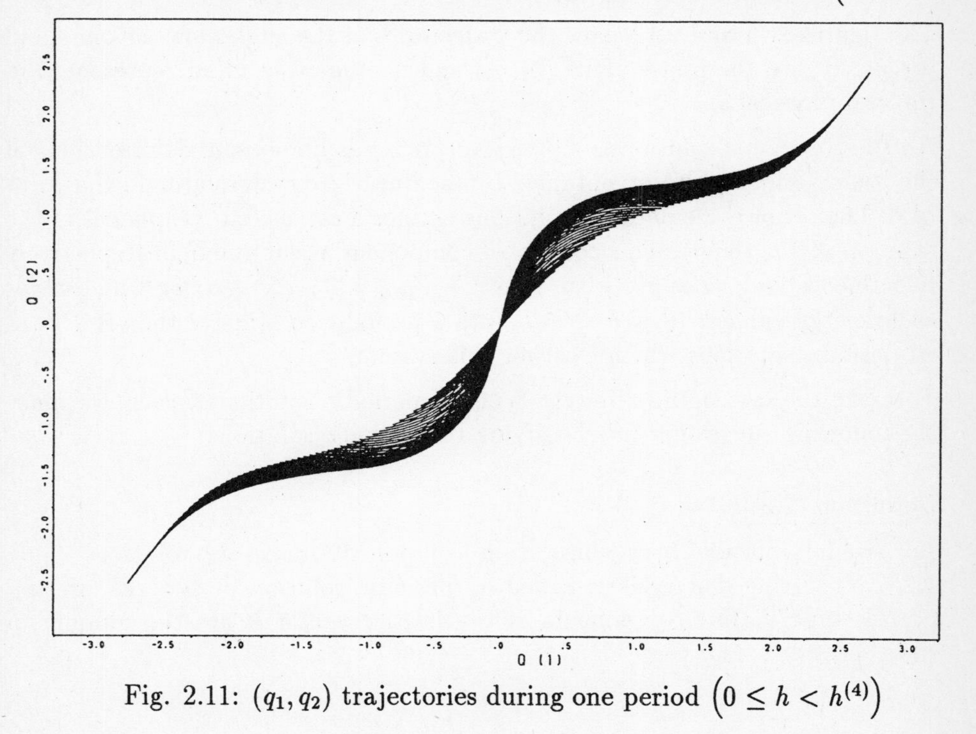

Fig. 2.11: (q_1, q_2) trajectories during one period $\left(0 \leq h < h^{(4)}\right)$

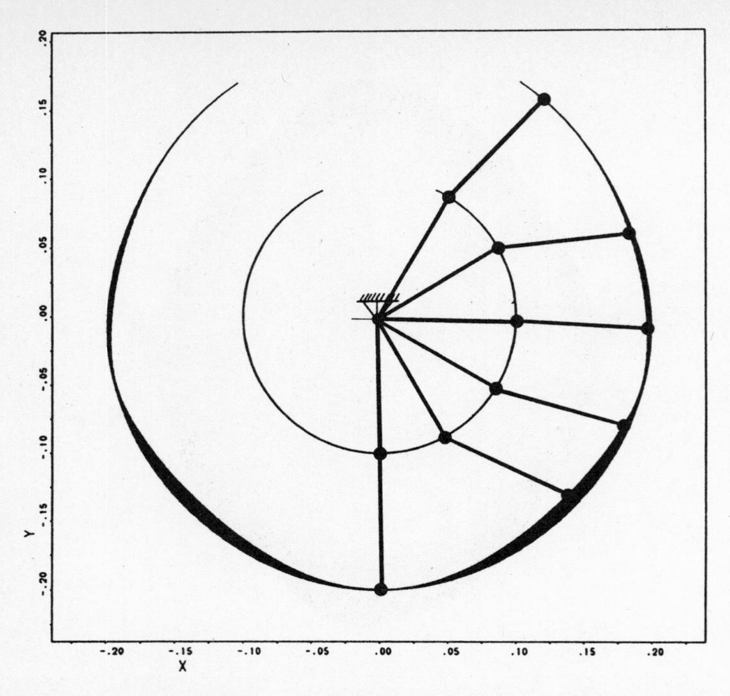

Fig. 2.12: periodic solutions represented in the real physical space $\left(0 \leq h < h^{(4)}\right)$

The figures 2.11 and 2.12 show the trajectories of the angles, on the one hand projected into the phase plane (q_1, q_2) and on the other hand represented in the real physical space.

All these periodic solutions may be interpreted as homotopic deformations of the basic periodic solution obtained by the linearized system around the point $\mathbf{q}^{(1)}$. The set of these periodic solutions defines a connected component of the type $(k_1, k_2) = (0\,,\,0)$. This connected component or submanifold respectively is defined about an appropriate function $Z_{(0,0)} : [0\,,\,h^{(4)}] \rightarrow Per$ which maps each energy value h $(0 \leq h < h^{(4)})$ into a periodic solution of the set Per of all periodic solutions which exist for this system.

To make the use of the different types of periodic solutions easier, we make the following suggestion for classifying the periodic solutions:

<u>Definition (2.2):</u>

A periodic solution $\mathbf{q}(t)$ whose j–th component makes k_j rotations $(j = 1, \ldots, f)$ during one cycle is called a "periodic solution of type $\mathbf{k}$" or simply "$\mathbf{k}$–type". The components of the integer vector $\mathbf{k}$ are the number of rotations $k_1, \ldots, k_f$.

○

N.B.: Definition (2.2) is not unique because there may be more periodic solutions which belong to the same energy value h.

(b) initial function:

$$Y_1^{(0)}(\tau) = A\cos(2\pi\tau) , \qquad Y_3^{(0)}(\tau) = -\omega A \sin(2\pi\tau) ,$$

$$Y_2^{(0)}(\tau) = -\sqrt{2}A\cdot\cos(2\pi\tau) , \qquad Y_4^{(0)}(\tau) = \tfrac{2\pi}{\omega}$$

with

$$\omega := 1.848\,\omega_0 , \qquad A := 0.1745 \quad (\hat{=}\, 10^o) .$$

The homotopy is constructed according to chapter (a). The result is a homotopic deformation of the second periodic basic solution of the linearized system around the point $\mathbf{q}^{(1)}$. The phase space pictures are similar to the ones shown in chapter (a). The figures 2.13 and 2.14 show the trajectories of the angles q_1 and q_2 represented in the phase space projection (q_1, q_2) and in the real physical space. As in case (a), the construction of the periodic solutions are generated from the fixed point $\mathbf{q}^{(1)}$. The homotopic deformation of the periodic solution ends at a homoclinic orbit which belongs to the energy value $h^{(4)}$. The range of the energy values of all periodic solutions which define the submanifold is $]0, h^{(4)}[$. The time of period converges for $h \to 0$ against the time of period of one of the periodic solutions of the linear part (cf. chapter 2.8).

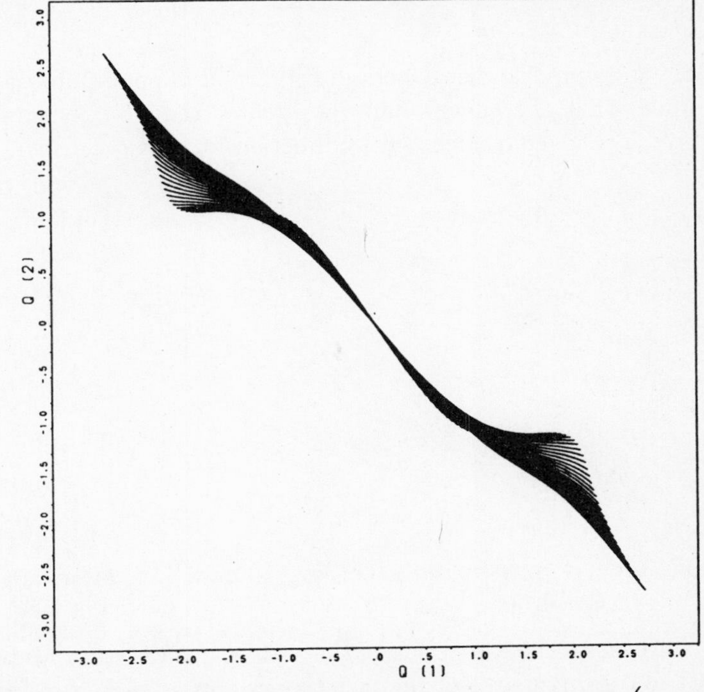

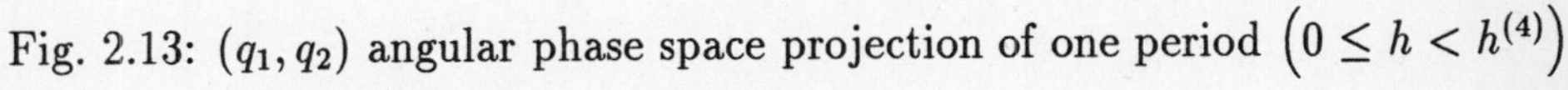
Fig. 2.13: (q_1, q_2) angular phase space projection of one period $\left(0 \le h < h^{(4)}\right)$

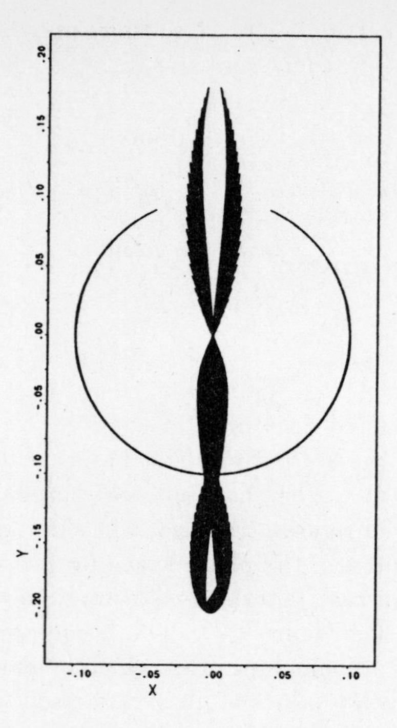

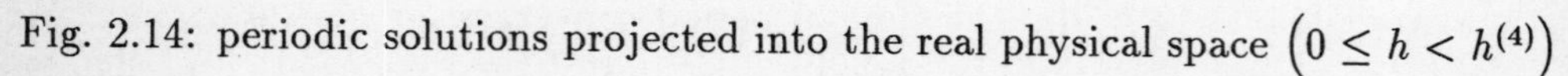
Fig. 2.14: periodic solutions projected into the real physical space $\left(0 \le h < h^{(4)}\right)$

<u>**2nd case:**</u> $k_1 = 0$; $k_2 = m$

m rotations of pendulum 2 and only periodic librations of pendulum 1 are considered. This case implies that the energy must be greater than $h^{(2)}$ because there is no rotation for $h < h^{(2)}$. A convenient initial function for this case is:

$$Y_1^{(0)}(\tau) = A \sin(2\pi\tau) , \qquad Y_3^{(0)}(\tau) = A\omega \cos(2\pi\tau) ,$$

$$Y_2^{(0)}(\tau) = -2\pi\, m\, \tau , \qquad Y_4^{(0)}(\tau) = \tfrac{2\pi}{\omega}$$

with

$$A := 0.1745 \quad (\hat{=}\, 10^o) , \qquad \omega := \frac{2\pi}{T^{(0)}} ; \quad T^{(0)} = m .$$

(a) <u>$m = 1$</u>

The homotopy is constructed according to case 1, that means the steps of energy are ($\Delta h = 0.1$).

Figure 2.15 shows the projection of the state space trajectory into the $(q_2, \dot{q}_2)$–plane. The turning points of the trajectories around $q_2 = \pi$ in fig. 2.15 will be investigated in chapter 2.7.

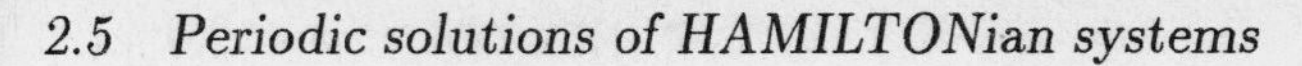

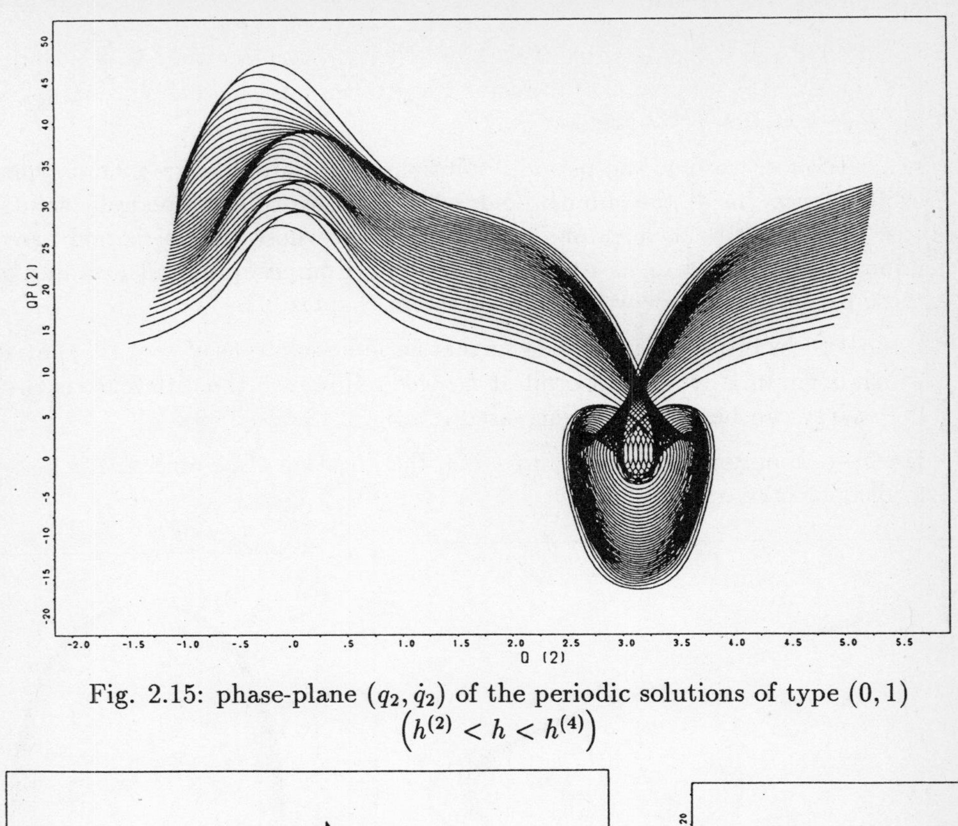

Fig. 2.15: phase-plane $(q_2, \dot{q}_2)$ of the periodic solutions of type $(0,1)$ $\left(h^{(2)} < h < h^{(4)}\right)$

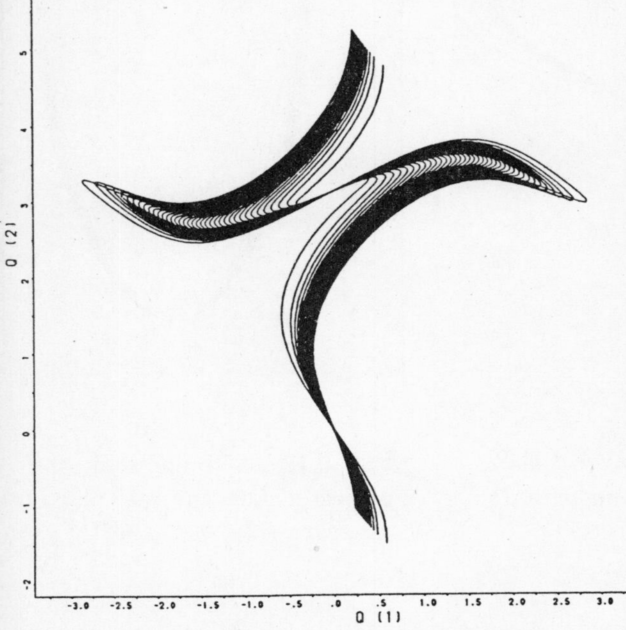

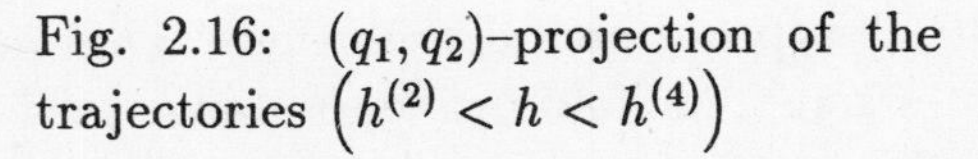
Fig. 2.16: (q_1, q_2)–projection of the trajectories $\left(h^{(2)} < h < h^{(4)}\right)$

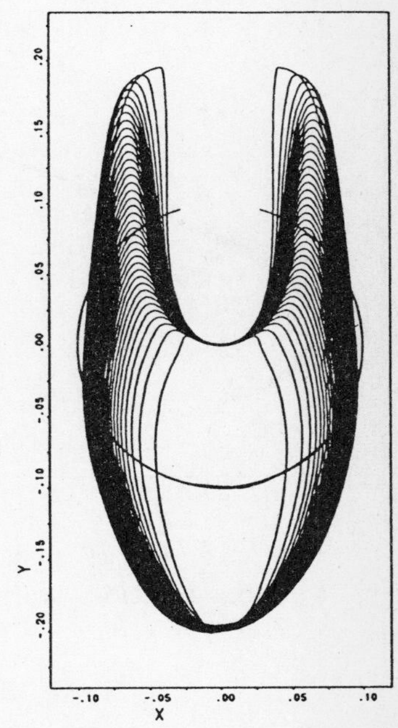

Fig. 2.17: periodic solutions projected into the real physical space $\left(h^{(2)} < h < h^{(4)}\right)$

Fig. 2.16 shows the projection of the periodic solutions into the (q_1, q_2)–plane, and fig. 2.17 the projection of the same trajectories into the real physical space for energy values $h^{(2)} < h < h^{(4)}$.

In contrary to case 1, the periodic solutions begin and end in a homoclinic orbit. Hence, inside the submanifold, which consists of these periodic solutions, there must be at least one periodic solution whose time of period has a minimum compared to its neighbor solutions. Computation and meaning of these periodic solutions will be considered in chapter 2.7.

Besides the solutions of type $(0,1)$ further periodic solutions of type $(0,1)$ exist which begin in a homoclinic orbit at $h = h^{(2)}$. However, the difference is that the energy can be arbitrarily increased (cf. fig. 2.18 and 2.19).

For $h \to \infty$ only pendulum 2 is rotating, the libration of pendulum 1 is going to disappear completely.

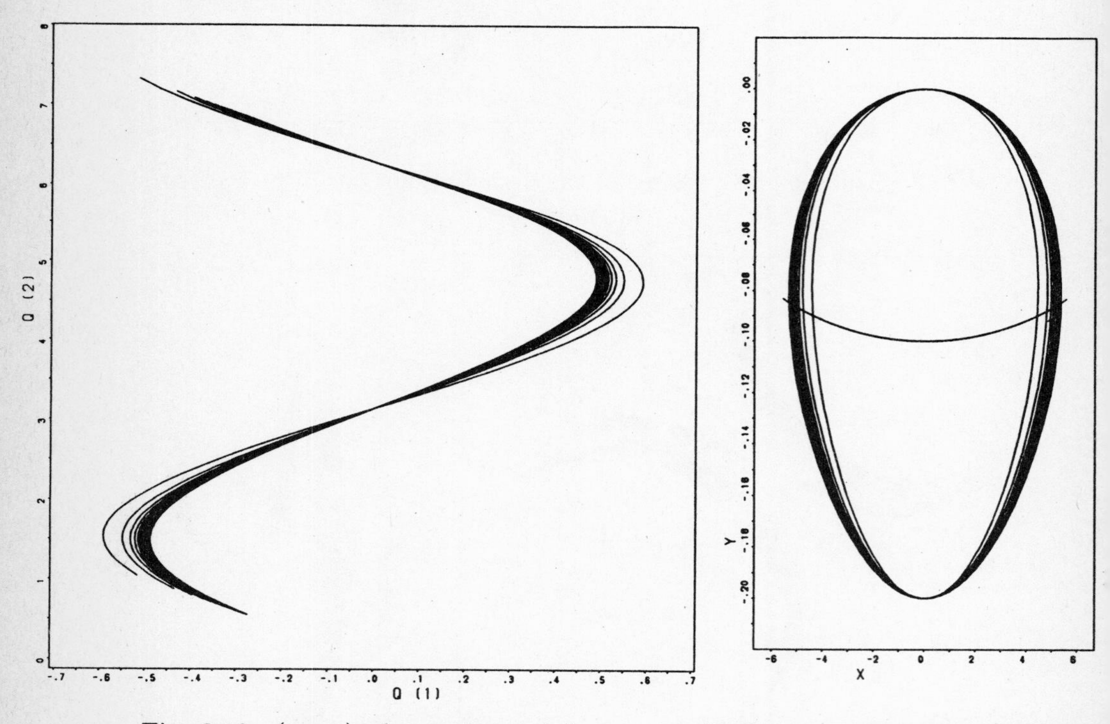

Fig. 2.18: (q_1, q_2) phase plane of the periodic solutions during one period $(h^{(2)} < h < h^{(4)})$

Fig. 2.19: periodic solutions projected into the real physical space $(h^{(2)} < h < h^{(4)})$

(b) <u>$m = 2$</u>

Similar to case (a) there are two $(0,2)$–types of periodic solutions. That means one set of periodic solutions defines a submanifold inside the range $]h^{(1)}, h^{(4)}[$

for the energy values h (cf. fig. 2.20 and 2.21). The other set of periodic solutions is a copy of the type $(0,1)$ inside the range $]h^{(2)}, \infty[$.

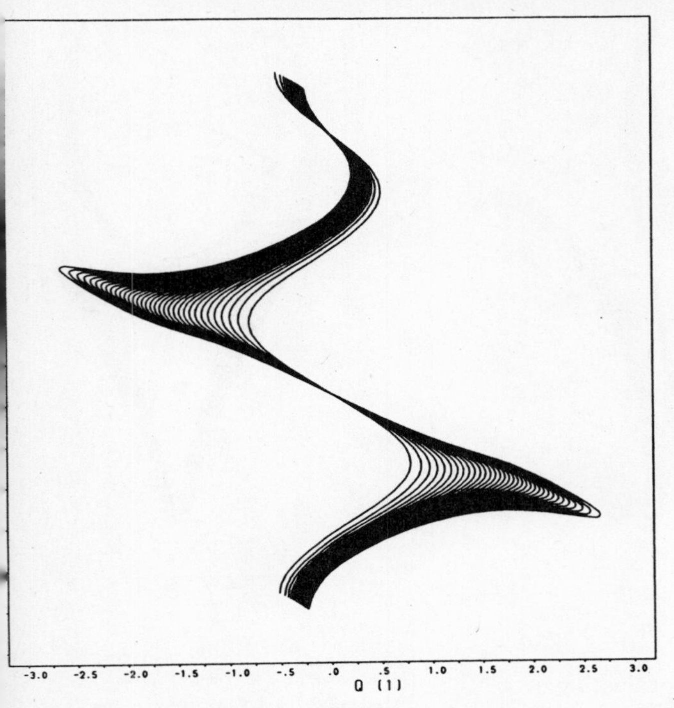

Fig. 2.20: (q_1, q_2) phase plane for periodic solutions of type $(0,2)$ $(h^{(2)} < h < h^{(4)})$

Fig. 2.21: periodic solutions of type $(0,2)$ projected into the real physical space $(h^{(2)} < h < h^{(4)})$

(c) $\underline{m \geq 3}$

All periodic solutions of that case are duplicates of the cases (a) and (b), since the number of librations during one period was not prescribed by the BVP.

<u>3rd case:</u> $k_1 = m \ , \quad k_2 = 0$

m rotations of pendulum 1 and only librations of pendulum 2 are considered. The initial function is chosen according to case 2. The figures 2.22 and 2.23 show the periodic solutions projected into the (q_1, q_2) phase plane and into the real physical space. The trajectories are qualitatively the same as in case 2 if q_1 is replaced by q_2 and vice versa. The range of energy is $]h^{(3)}, h^{(4)}[$ and the submanifold consisting of all periodic solutions of type $(m, 0)$ is limited by two homoclinic orbits.

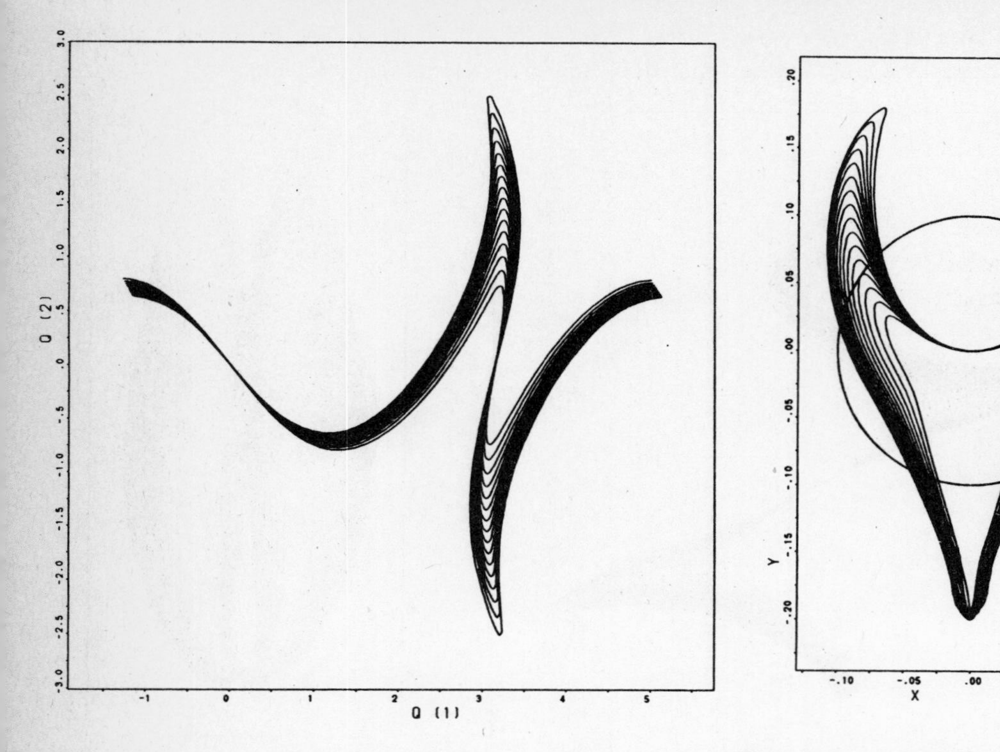

Fig. 2.22: (q_1, q_2) phase plane projection of the trajectories $(h^{(3)} < h < h^{(4)})$

Fig. 2.23: periodic solutions projected into the real physical space $(h^{(3)} < h < h^{(4)})$

4th case: $k_1 = m \ , \quad k_2 = n$

The energy value h must be bigger than $h^{(4)}$ because pendulum 1 rotates m–times and pendulum 2 rotates n–times during one period. In that case a metric on the configuration space T^2 without singularities can be constructed. Based on that metric it can be shown that for each $m, n \in \mathbb{Z} - \{0\}$ a closed geodetic or periodic function respectively exists (cf. [ARNOLD 1984], [KLINGENBERG 1978]).

Each submanifold of periodic solutions of type (m, n) is limited by a homoclinic orbit (energy value $h = h^{(6)}$). The range of energy values is $]h^{(6)}, \infty[$ and we obtain $\lim_{h \to \infty} T(h) = 0$.

For the numerical computation of periodic solutions of type (m, n) we chose the initial function:

$$Y_1^{(0)} := 2m\pi\tau \ , \qquad Y_3^{(0)} := m \ ,$$

$$Y_2^{(0)} := 2n\pi\tau \ , \qquad Y_4^{(0)} := n$$

with $m\,,\ n \in \mathbb{Z} - \{0\}$.

The following figures show the results for some combinations of m and n:

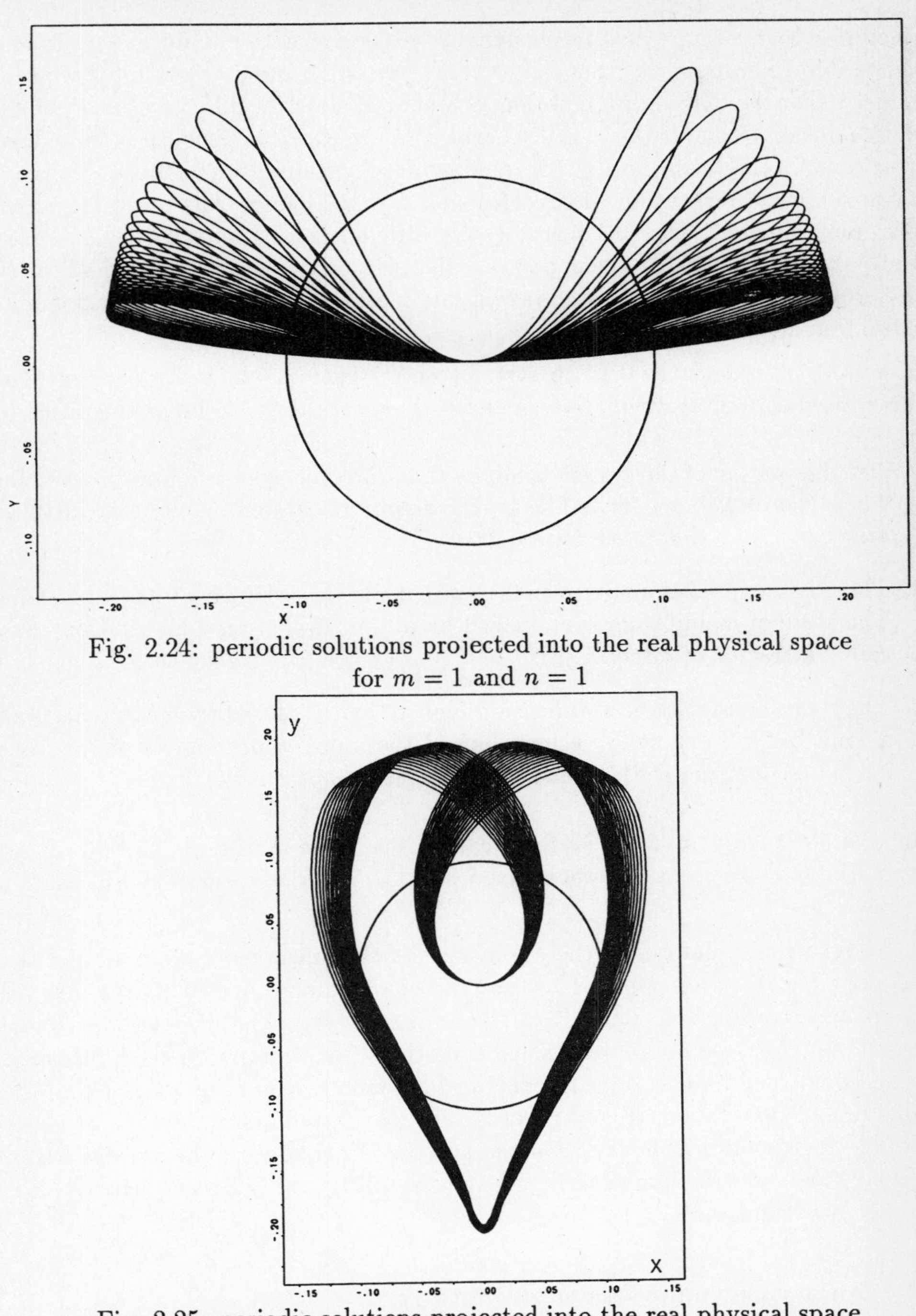

Fig. 2.24: periodic solutions projected into the real physical space for $m = 1$ and $n = 1$

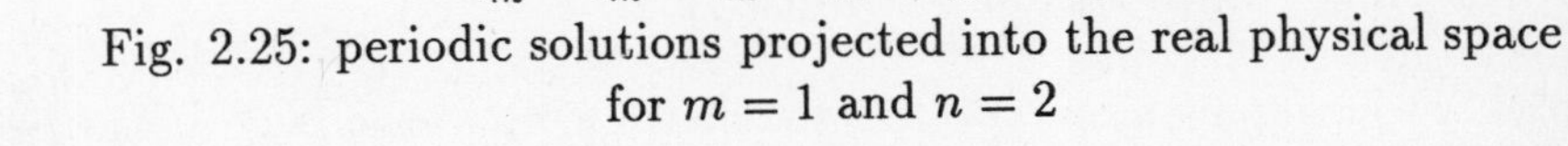

Fig. 2.25: periodic solutions projected into the real physical space for $m = 1$ and $n = 2$

2.6 Periodic solutions of dissipative and excited systems

Trajectories $\Phi_x(t)$ of pure dissipative dynamical systems converge for $t \to \infty$ to one of the stable fixed points of the field $\mathbf{f}$. If there are more stable fixed points of the field $\mathbf{f}$ then we obtain different domains of attraction for different fixed points. The domains of attraction are usually derived by separatrices (smooth manifolds of dimension $< 2f$). The eigenvalues of the linearized systems around the stable fixed point are – according to the theory of nomal forms (chapter 2.3) – not resonant. Hence, pure dissipative systems are always diffeomorph to one of the linearized systems around the stable fixed points. Because of this reason, the qualitative behavior of such systems is easy to investigate if all fixed points and their domains of attraction are known.

This is not the case in HOPF'ian systems or in systems which are outside- and inside-excited as well as dissipative. At least three cases have to be distinguished:

1. The dissipation of the system is higher than the flow of energy into the system. The dynamical behavior is like a pure dissipative system. Only attractors of dimension 0 (stable fixed points) exist.

2. There is an equilibrium between dissipation and flow of energy into the system. The state of equilibrium is expressed by a limit cycle (periodic solution) also called attractor of dimension 1.

3. There are attractors of a dimension higher than 1 caused by a bifurcation of a limit cycle if one of the eigenvalues of the monodromy matrix of the limit cycle has the form $e^{2\pi i\nu}$ with $\nu \in \mathbb{R} - \mathbb{Q}$ (see chapter 2.8).

If the flow of energy into the system reaches the value of the dissipation of the system a bifurcation from the stable fixed point to a periodic solution will usually take place.

At the bifurcation point one of the eigenvalues of the linearized system around the fixed point becomes resonant. The type of the singularity (fixed point) may be investigated according to chapter 2.4. The limit cycle after the bifurcation is stable provided that the fixed point was stable before the bifurcation. After the bifurcation the fixed point will be unstable. Bifurcations are mostly caused by variation of the parameters of the system. Fixed points, limit cycles and their stability behavior determine the dynamical behavior of the system. Therefore the knowledge about the fixed points and/or limit cycles and their stability and bifurcation behavior is the most important part.

2.6.1 Numerical computation of limit cycles

One possibility to compute a limit cycle is to determine the fixed point of the field $\mathbf{f}$ and choose a bifurcation parameter $\mathbf{p}$, which controls the dissipation in the system.

The next step is the computation of the initial value problem whose initial value lies in the neighborhood of the fixed point. The parameter corresponds to an initial value problem which belongs to an asymptotically stable solution. In the second step, the parameter $\mathbf{p}$ will be varied untill the first limit cycle occurs as a solution of the initial value problem. However, the method is restricted to stable limit cycles. A more efficient and general method is to use the bifurcation parameter $\mathbf{p}$ as a homotopy parameter and compute the limit cycle by a boundary value problem (BVP) of the form

$$\begin{aligned} \dot{\mathbf{x}} &= \mathbf{f}(\mathbf{x},\mathbf{p}) \ , \\ \mathbf{x}(t) &= \mathbf{x}(0) + 2\mathbf{k}\pi \ . \end{aligned} \tag{2.108}$$

In HAMILTONian systems an additional algebraic constraint has to be taken into account (cf. chapter 2.5). Methods to solve the BVP (2.108) have been proposed by [SEYDEL 1981] and [DEUFLHARD 1984] for the case $\mathbf{k} = \mathbf{0}$. Dynamical systems are always represented by an ODE of second order. Therefore it is not necessary to use algorithms proposed by DEUFLHARD since the boundary condition (equation (2.97)) can be used in any case to get rid of the problem of invariance of translation[10]. If the interval $[0,T]$ is tansformed onto the interval $[0,1]$ according to equation (2.105) and if the definitions

$$\tau := \frac{t}{T} \ ; \quad \left(' := \frac{d}{dt} \right) \ , \tag{2.109}$$

$$\mathbf{y}(\tau) := \begin{bmatrix} \mathbf{x}(t) \\ T \end{bmatrix} \in \mathbb{R}^{2f+1} \ , \tag{2.110}$$

$$\mathbf{r}(\mathbf{y}(0)\,,\mathbf{y}(1)) := \begin{bmatrix} y_1(1) - y_1(0) \\ \vdots \\ y_{2f}(1) - y_{2f}(0) \\ \\ y_{f+i}(0) \cdot y_{2f+1}(0) - 2k_i\pi \end{bmatrix} \ , \tag{2.111}$$

$$(i \in \{1,\ldots,f\} \text{ arbitrary!})$$

are employed, a completely formulated BVP results:

$$\mathbf{y}' = \begin{bmatrix} \mathbf{f}(\mathbf{y},\mathbf{p}) \cdot y_{2f+1} \\ 0 \end{bmatrix} \ , \tag{2.112}$$

---

[10]The mean value theorem can also be applied to general ODE's to obtain the necessary boundary value in order to eliminate the problem of invariance of translation. However, in general the boundary condition will be non-linear and is according to equation (2.97):

$$f_j\,(\mathbf{x}(0),\,\mathbf{p}) = \frac{2k_j\pi}{T} \qquad \text{for some} \quad j \in \{1,\ldots,f\} \ .$$

$$\mathbf{0} \;=\; \mathbf{r}(\mathbf{y}(0)\,,\,\mathbf{y}(1))\;. \tag{2.113}$$

2.6.2 Example: Wheel set of a railway vehicle

A nontrivial example for a HOPF-bifurcation to limit cycle motions is the wheel set shown in fig. 2.1.

(a) Change of the definition of coordinate frame K

We assume that the coordinate frame K is not body fixed and the e_x^K-axis always points to the direction of the rotational axis of the wheel set. For an observer, who is fixed in the K-system, the wheel set appears to have the angular velocity $\dot{\beta}_0 = -\frac{v_0}{r_0}$ if the wheel set moves straight with velocity v_0. In that case, the tensor of the mass of inertia represented in the frame K is constant. Furthermore, the equations of motion are not modified.

(b) Model of rolling friction

A model of rolling friction which defines the contact forces $\mathbf{F}_L$ and $\mathbf{F}_R$ as a function of the state $(\mathbf{r}, \mathbf{T}, \mathbf{v}, \boldsymbol{\omega})$ of the system is necessary. A corresponding model based on the theorie of KALKER, can be found in [JASCHINSKI 1987], [JOCHIM 1987] or [SIMEON 1988]: If N is the component of the contact force between wheel and rail orthogonal to the tangent plane at the point of contact – i.e. if N is the component into the direction e_z^L or e_z^R respectively (cf. fig. 2.1) – then the contact forces into the direction e_x^L or e_x^R respectively are

$$W_x = \mu\, N\, \tanh\left(k_x \cdot \frac{s_x}{N}\right)$$

and the contact forces into the direction e_y^L or e_y^R respectively are

$$W_y = -\mu\, N\, \tanh\left(\frac{k_y s_y + k_z s_z}{N}\right)\;.$$

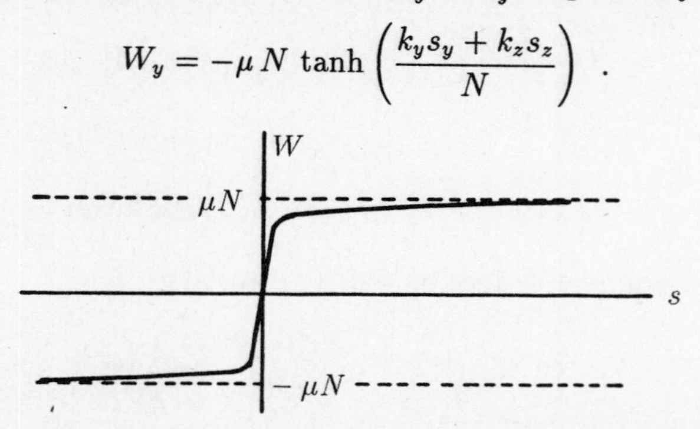

Fig. 2.26: drag W of rolling friction as a function of the slip s

Represented in the frame L or R respectively, the contact force has the components

$$(W_x\,,\, W_y\,,\, N)\;.$$

The slip-vector $\mathbf{s} := (s_x, s_y, s_z)$ which is defined by

$$\mathbf{s} := \frac{1}{v_0} \begin{bmatrix} v_{rel,x} \\ v_{rel,y} \\ \omega_{rel,z} \end{bmatrix} = \begin{bmatrix} \text{longitudinal slip} \\ \text{transversal slip} \\ \text{spin} \end{bmatrix} .$$

$\mathbf{v}_{rel}$ ($\boldsymbol{\omega}_{rel}$) is the relative velocity (relative angular velocity) between wheel and rail at the contact point $C_L(C_R)$ represented in the frame L (R).

The constant vector $\mathbf{k} := (k_x, k_y, k_z)$ contains the shear modul G, the KALKER-coefficients C_{ik}, the friction coefficient μ and some geometrical coefficients of the contact ellipsoid.

(c) Stationary solution

Based on the model of rolling friction the functions

$$\begin{aligned} x(t) &= 0 \quad , & \alpha(t) &= 0 \; , \\ y(t) &= 0 \quad , & \beta(t) &= \dot{\beta}_0 t \; , \\ z(t) &= 0 \quad , & \gamma(t) &= 0 \end{aligned}$$

represent a (stationary) solution or a singular point respectively of the equations of motion.

The disturbances of the stationary state are described by the deviations of position x, y, z of the origin of frame K with respect to frame I and by the angular deviations $\alpha, \Delta\beta, \gamma$ of the unit vectors of frame K with respect to the unit vectors of frame I.

The normal force N has the same amount on the left hand side as well as on the right hand side in the stationary case. N is the solution of the non-linear algebraic equation

$$g(N) := \left(\cos\delta - \mu \sin\delta \cdot th \left[\frac{k_z \cdot \sin\delta}{r_0 \cdot N} \right] \right) N - \frac{mg}{2} = 0 \; .$$

This equation has only one solution in the range $0 \le \delta < \frac{\pi}{2}$.

(d) Position vectors of coordinate transformations

From the practical point of view it is appropriate to assume that the angles $\alpha, \Delta\beta$ (in the following β instead of $\Delta\beta$) and γ are small. In this case the geometrical constraints can be simplified to

$$\mathbf{h}(\mathbf{r}, \mathbf{T}) = \begin{bmatrix} \gamma \left[(a - x) - (r_0 + z) tg\delta \right] + x tg\delta - z \\ \\ \gamma \left[(a + x) - (r_0 + z) tg\delta \right] + x tg\delta + z \end{bmatrix} = \mathbf{0} \; ,$$

or, expressed in coordinates x and z:

$$\begin{bmatrix} x \\ z \end{bmatrix} = \left(\frac{a - r_0 tg\delta}{tg\delta}\right) \begin{bmatrix} -\gamma \\ \gamma^2 \end{bmatrix} .$$

I.e. for small angles γ the deviations x and z are small too.

Hence, the position vectors from the center of mass S of the wheel set to the contact point C_L or C_R respectively expressed in the frame K have the form:

$$\mathbf{r}_L = \begin{bmatrix} -a + \frac{1}{tg\delta}(\gamma a + z) \\ \alpha r_0 tg\delta - \beta r_0 \\ -\gamma a - r_0 - z \end{bmatrix} , \qquad \mathbf{r}_R = \begin{bmatrix} a + \frac{1}{tg\delta}(\gamma a - z) \\ -\alpha r_0 tg\delta - \beta r_0 \\ \gamma a - r_0 - z \end{bmatrix} .$$

If $K_B : V \to \mathbb{R}^3$, $\mathbf{u} \mapsto {}_B\mathbf{u}$ is the "coordinate mapping" of any vector $\mathbf{u}$ of the 3-dimensional vector space V into frame B of V, then the relations

$$\begin{aligned} K_I &= \mathbf{T}_K \circ K_K , \\ K_K &= \mathbf{T}_L \circ K_L , \\ K_K &= \mathbf{T}_R \circ K_R \end{aligned}$$

define the matrices $\mathbf{T}_K$, $\mathbf{T}_L$ and $\mathbf{T}_R$ by:

$$\mathbf{T}_K = \mathbf{E} + \tilde{\boldsymbol{\varphi}} \; ; \quad \boldsymbol{\varphi} := (\beta, \gamma, \alpha)^T ,$$

$$\mathbf{T}_L = \begin{bmatrix} -\alpha & c\delta & s\delta \\ -1 & -(\alpha c\delta + \beta s\delta) & (\beta c\delta - \alpha s\delta) \\ \beta & -s\delta & c\delta \end{bmatrix} ,$$

$$\mathbf{T}_R = \begin{bmatrix} -\alpha & c\delta & -s\delta \\ -1 & (\beta s\delta - \alpha c\delta) & (\alpha s\delta + \beta c\delta) \\ \beta & s\delta & c\delta \end{bmatrix}$$

with $c\delta := \cos\delta$ and $s\delta := \sin\delta$.

(e) Forces and torques

The wheel set is elastically mounted in the suspension (cf. fig. 2.1), and the inertial frame defines the principal axes for the linear stiffnesses of the bearings. The equivalent pair of force and torque $(\mathbf{F}, \mathbf{M})$, that acts from the suspension to the wheel set, is represented in frame I or K respectively by:

$$
\begin{aligned}
{}_I\mathbf{F} &= {}_I\mathbf{G} - \begin{bmatrix} c_x & 0 & 0 \\ 0 & c_y & 0 \\ 0 & 0 & c_z \end{bmatrix} \begin{bmatrix} x \\ y \\ z \end{bmatrix} , \\
{}_K\mathbf{M} &= -\mathbf{T}_K^T \cdot \begin{bmatrix} 0 & 0 & 0 \\ 0 & c_\gamma & 0 \\ 0 & 0 & c_\alpha \end{bmatrix} \begin{bmatrix} \beta \\ \gamma \\ \alpha \end{bmatrix} .
\end{aligned}
$$

The vector ${}_I\mathbf{G}$ takes care of the gravitational part. If the wheel set runs straight forward, then ${}_I\mathbf{G} = (0, 0, -mg)$. $c_x, \ldots, c_\gamma$ are the stiffnesses of the mounting elements in $x, \ldots, \gamma$-direction. The β-direction is free.

(f) Equations of motion

The motion of the system is described by a differential algebraic system (DAE)

ODE	$\dot{\mathbf{r}} = \mathbf{v}$ $\dot{\varphi} = \omega - \omega_0$ $\dot{\mathbf{v}} = \frac{1}{m}\left[{}_I\mathbf{F} + {}_I\mathbf{F}_L + {}_I\mathbf{F}_R\right]$ $\dot{\omega} = \mathbf{J}^{-1}\left[{}_K\mathbf{M} - \tilde{\omega}\mathbf{J}\omega + {}_K\mathbf{r}_L \times {}_K\mathbf{F}_L + {}_K\mathbf{r}_R \times {}_K\mathbf{F}_R\right]$
AC	$\mathbf{0} = \begin{bmatrix} x \\ z \end{bmatrix} + \frac{1}{\Gamma}\begin{bmatrix} \gamma \\ -\gamma^2 \end{bmatrix}$

The following definitions are made:

$$\begin{aligned}
\boldsymbol{\omega}_0 &:= \left(\dot{\beta}_0\,,\,0\,,\,0\right)^{\cdot}\,, \\
\boldsymbol{\omega} &:= {}_K\boldsymbol{\omega}\,, \\
\mathbf{v} &:= {}_I\mathbf{v}\,, \\
\mathbf{J} &:= {}_K\mathbf{J}_S = diag\,\{I_x\,,\,I_y\,,\,I_z\}\,, \\
\mathbf{r} &:= {}_I\mathbf{r} = (x\,,\,y\,,\,z)^T\,, \\
\Gamma &:= \left(\frac{tg\delta}{a - r_0 tg\delta}\right)\,.
\end{aligned}$$

Because of $z = o(\gamma^2)$ the algebraic constraint (AC) can be expressed by

$$\mathbf{h}(\mathbf{r},\boldsymbol{\varphi}) = \begin{bmatrix} x + \frac{\gamma}{\Gamma} \\ z \end{bmatrix} = \mathbf{0}\,.$$

Hence, the DAE is

$$\begin{aligned}
\dot{\mathbf{x}} &= \mathbf{f}(\mathbf{x},\mathbf{p},\boldsymbol{\lambda})\,, \\
\mathbf{0} &= \mathbf{h}(\mathbf{x},\mathbf{p})
\end{aligned}$$

with

$$\begin{aligned}
\mathbf{x} &:= \left(\mathbf{r}^T,\,\boldsymbol{\varphi}^T,\,\mathbf{v}^T,\,\boldsymbol{\omega}^T\right)\,, \\
\mathbf{p} &:= (a\,,\,r_0\,,\,v_0\,,\,\delta\,,\,\mu\,,\,c_x\,,\,c_y\,,\ldots)\,, \\
\boldsymbol{\lambda} &:= (N_L\,,\,N_R)\,.
\end{aligned}$$

N_L and N_R are the normal forces at the left and right contact point. N_L and N_R are not yet known. To eliminate N_L and N_R, the second derivative of the algebraic constraint has to be employed:

$$\Rightarrow \quad \begin{bmatrix} \ddot{x} + \frac{1}{\Gamma}\ddot{\gamma} \\ \ddot{z} \end{bmatrix} = \mathbf{0}\,,$$

or,

$$\mathbf{g}(\mathbf{x},\mathbf{p},\boldsymbol{\lambda}) := \begin{bmatrix} \dot{v}_1(\mathbf{x},\mathbf{p},\boldsymbol{\lambda}) + \frac{1}{\Gamma}\dot{\omega}_2(\mathbf{x},\mathbf{p},\boldsymbol{\lambda}) \\ \dot{v}_3(\mathbf{x},\mathbf{p},\boldsymbol{\lambda}) \end{bmatrix} = 0\,.$$

The coordinates x and z that occur in the ODE will be eliminated by using the algebraic constraint, that is:

$$x = -\tfrac{1}{\Gamma}\gamma \quad , \qquad z = 0 \ ,$$

$$\dot{x} = -\tfrac{1}{\Gamma}\dot{\gamma} \quad , \qquad \dot{z} = 0 \ .$$

The new variables are

$$\mathbf{Y} \ := \ (y\,,\beta\,,\gamma\,,\alpha\,,v_2\,,\omega_1\,,\omega_2\,,\omega_3)^T \ .$$

$$\Rightarrow \qquad \dot{\mathbf{Y}} \ = \ \begin{bmatrix} v_2 \\ \omega_1 \\ \omega_2 \\ \omega_3 \\ \dot{v}_2 \\ \dot{\omega}_1 \\ \dot{\omega}_2 \\ \dot{\omega}_3 \end{bmatrix} \ =: \hat{\mathbf{f}}(\mathbf{Y},\mathbf{p},\boldsymbol{\lambda}) \ .$$

In the same way x, z and $\dot{x}, \dot{z}$ are eliminated in equation $\mathbf{g}(\mathbf{x},\boldsymbol{\lambda}) = \mathbf{0}$.

$$\Rightarrow \qquad \hat{\mathbf{g}}(\mathbf{Y},\mathbf{p},\boldsymbol{\lambda}) := \mathbf{g}(\mathbf{x}(\mathbf{Y})\,,\,\mathbf{p}\,,\,\boldsymbol{\lambda}) = \mathbf{0} \ .$$

If $\hat{\mathbf{g}}$ is numerically solved with respect to $\boldsymbol{\lambda}$, for instance $\boldsymbol{\lambda} = \boldsymbol{\psi}(\mathbf{Y},\mathbf{p})$, the standardform according to equation (2.5) is:

$$\dot{\mathbf{Y}} = \hat{\mathbf{f}}(\mathbf{Y}\,,\,\mathbf{p}\,,\,\boldsymbol{\psi}(\mathbf{Y},\mathbf{p})) =: \mathbf{F}(\mathbf{Y},\mathbf{p}) \ .$$

(g) Computation of periodic solutions (limit cycles)

$\mathbf{Y}_0(t) := (0\,,0\,,0\,,0\,,0\,,\dot{\beta}_0\,,0\,,0)^T$ is a singular point of the vector field $\mathbf{F}$ (cf. part (c)). In the neighborhood of that fixed point the periodic solutions of the non-linear system are diffeomorph to the periodic eigenmodes (if existent) of the linearized system around the fixed point (cf. chapter 2.3). It will be shown in chapter 2.8.2 that the solution $\mathbf{Y}_0(t)$ bifurcates to a periodic solution at $v_0^* \approx 19.0\frac{m}{s}$. At the bifurcation point the system is resonant, that is the JACOBIan $D\mathbf{F}(\mathbf{Y}_0,\mathbf{p}^*)$ has a pure imaginary pair of eigenvalues. If $\lambda,\overline{\lambda} = \pm i\omega$ is that pair, then the periodic eigenmode of the linearized system around $\mathbf{Y}_0$ is

$$\mathbf{Y}^{(0)}(t) := C \cdot (\mathbf{u}\cos\omega t + \mathbf{v}\sin\omega t) \ ,$$

where $\mathbf{u} + i\mathbf{v}$ is the eigenvector that belongs to the eigenvalue λ.

If $\mathbf{Y}^{(0)}(t)$ is used as initial function to compute the limit cycles for velocities $v_0 > v_0^*$ the solutions for the angles α and γ, shown in the figures 2.27 to

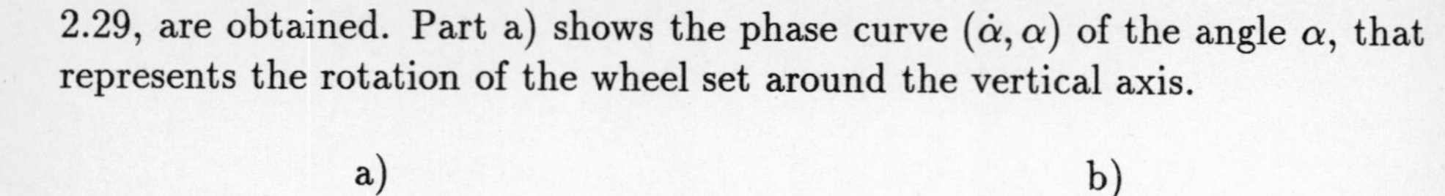

2.29, are obtained. Part a) shows the phase curve $(\dot{\alpha}, \alpha)$ of the angle α, that represents the rotation of the wheel set around the vertical axis.

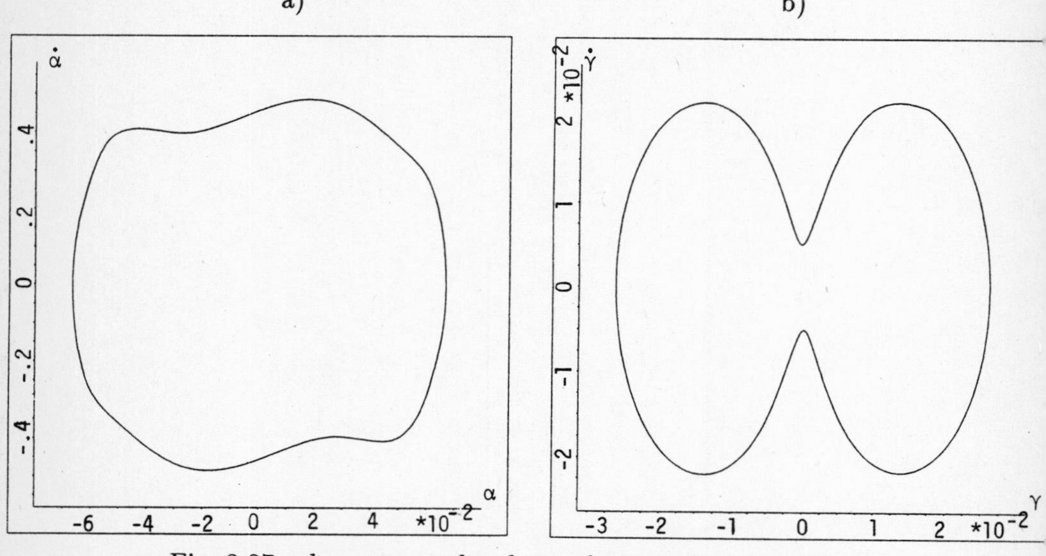

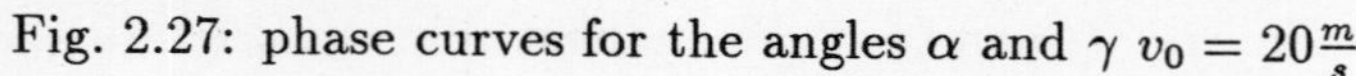

Fig. 2.27: phase curves for the angles α and γ $v_0 = 20\frac{m}{s}$

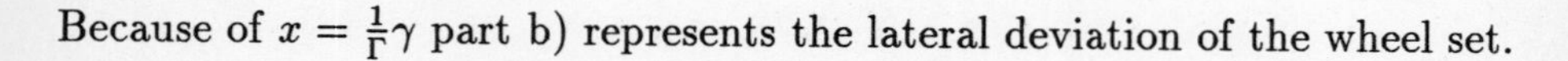

Because of $x = \frac{1}{\Gamma}\gamma$ part b) represents the lateral deviation of the wheel set.

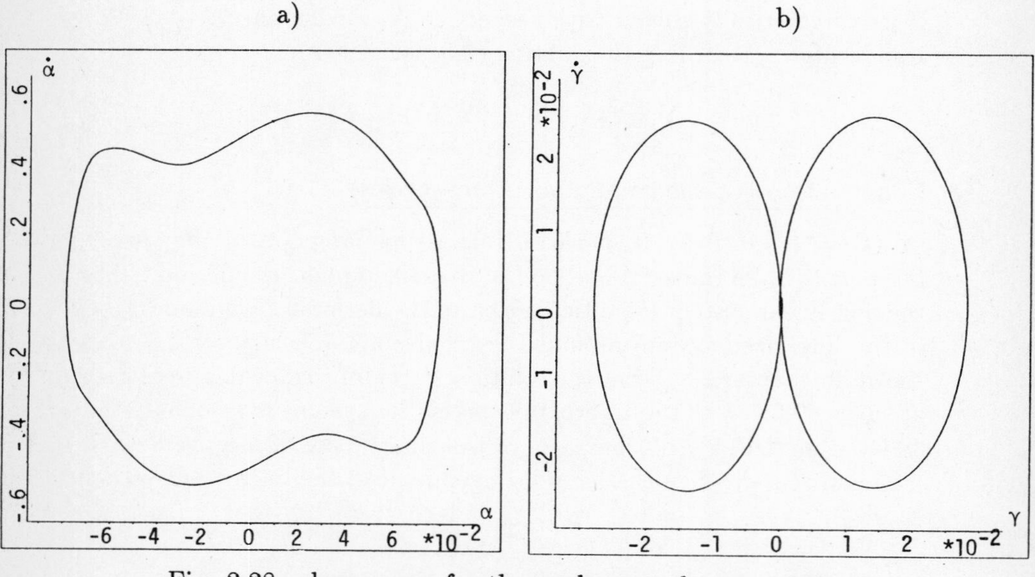

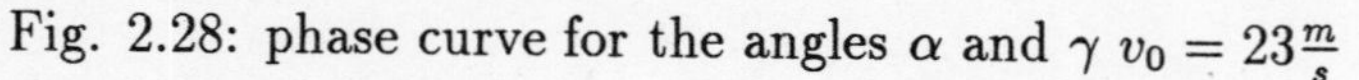

Fig. 2.28: phase curve for the angles α and γ $v_0 = 23\frac{m}{s}$

At $v_0 \approx 23\frac{m}{s}$ a "turning point" occurs (cf. chapter 2.7) that causes an additional lateral deviation around the center line of the rail.

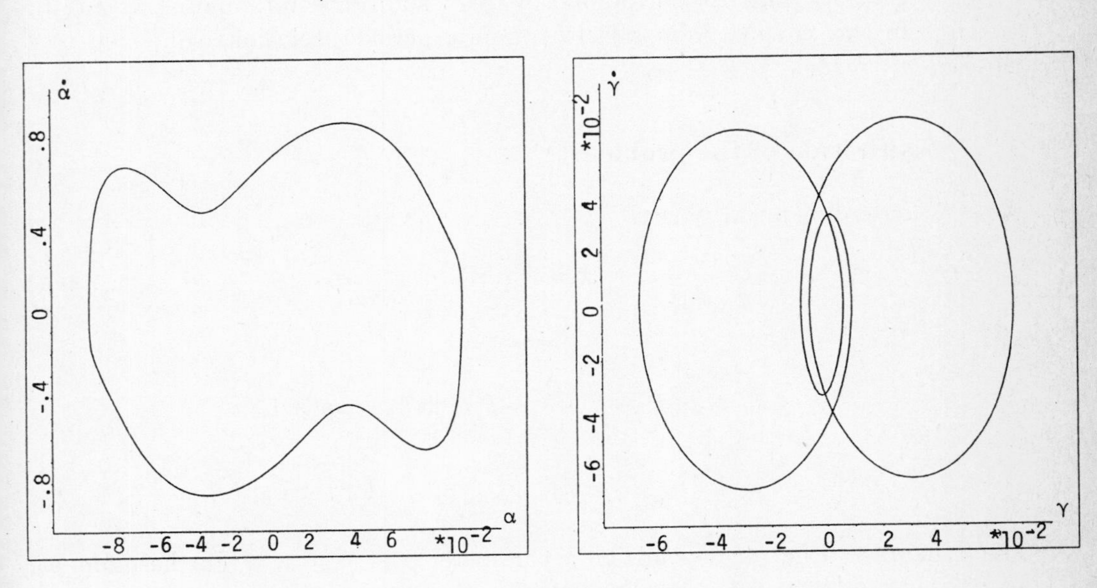

Fig. 2.29: phase curve of the angles α and γ $v_0 = 28\frac{m}{s}$

A modification of the dynamical model, used here, can be found in [MEIJAARD, DE PATER 1989]. Instead of the cone profile a cylinder with rim is used. The mathematical model is simpler than the one investigated above. However, the system contains discontinuities that stem from the impacts between rail and rim.

2.7 Periodic solutions with additional properties

As shown in chapter 2.5.3 case 2, periodic solutions of dynamical systems whose time T of period is minimal inside the range of its connected component may exist. Periodic solutions with special properties carry additional information about the system. In this chapter proposals to compute periodic solutions subjected to a criterion are made.

2.7.1 Definition of the problem

We consider a dynamical system

$$\dot{\mathbf{x}} = \mathbf{f}(\mathbf{x}, \mathbf{p}) \tag{2.114}$$

with

(1) $$\mathbf{f} : TM \times P \to TM \ ; \quad P \subset \mathbb{R}^m \ ; \quad m \geq 1$$

(2) $$\mathbf{f}(\ldots, x_i + 2\pi, \ldots) = \mathbf{f}(\ldots, x_i, \ldots) \quad \text{for several } i \ ,$$

The dynamical behavior depends on a parameter $\mathbf{p} \in P \subset \mathbb{R}^m$. Furthermore, we assume that for all $\mathbf{p} \in P$ a locally isolated periodic solution for any initial value $\boldsymbol{\xi}$ exists:

$$\begin{aligned} &\Phi_\xi(\cdot, \mathbf{p}) : \mathbb{R} \to TM \ , \\ &\Phi_\xi(T_\mathbf{p}, \mathbf{p}) = \Phi_\xi(0, \mathbf{p}) + 2\mathbf{k}\pi \ , \quad \mathbf{k} \in \mathbb{Z}^n \ . \end{aligned} \tag{2.115}$$

The time of period T usually depends on $\mathbf{p}$. Finally we want to assume that a non-linear functional, such as an integral criterion

$$F : C^1(\mathbb{R}, TM) \times P \to \mathbb{R} \ , \qquad (\mathbf{x}, \mathbf{p}) \mapsto \int_0^T g(\mathbf{x}, \mathbf{p}) dt \tag{2.116}$$

is given.

The criterion is used to locate periodic solutions with special properties inside their connected component. $g : TM \times P \to \mathbb{R}$ is a given cost function.

▽

Example 2.17: (integral criterion with respect to time of period T)

$$g(\mathbf{x}, \mathbf{p}) \ := \ 1 \ .$$

$$\Rightarrow \quad F(\mathbf{x}, \mathbf{p}) \ = \ \int_o^T 1 dt \ = \ T \ .$$

△

∇

Example 2.18: (integral criterion with respect to divergence of the vector field)

$$\begin{aligned} g(\mathbf{x},\mathbf{p}) &:= div\,(\mathbf{f}(\mathbf{x},\mathbf{p}))\ . \\ \Rightarrow\quad F(\mathbf{x},\mathbf{p}) &= \textstyle\int_0^T div\,(\mathbf{f}(\mathbf{x},\mathbf{p}))\,dt\ . \end{aligned}$$

$\triangle$

The criterion in example 2.18 defines F to be a measure of stability of the periodic solutions or of the limit cycle respectively. This result is summarized in theorem (2.3.)

Theorem (2.3):

If $\Phi_\xi(\cdot,\mathbf{p})$ is a periodic solution of the dynamical system $\dot{\mathbf{x}} = \mathbf{f}(\mathbf{x},\mathbf{p})$ and the eigenvalues λ_i of the monodromy matrix of the linearized system around the periodic solution $\Phi_\xi(\cdot,\mathbf{p})$ can be expressed by

$$\lambda_i = e^{\mu_i} \qquad (i = 1,\ldots,2f) \tag{2.117}$$

(cf. chapter 2.8), then

$$\sum_{i=1}^{2f} \mu_i = \int_0^T div\,\mathbf{f}(\mathbf{x}(t),\mathbf{p})\,dt\ . \tag{2.118}$$

Proof:

Assume that $\mathbf{W}(t)$ is the WRONSKIan matrix of the linearized system

$$\dot{\mathbf{y}} = D\mathbf{f}\,(\Phi_\xi(t,\mathbf{p}),\mathbf{p})\,\mathbf{y}$$

around $\Phi_\xi(\cdot,\mathbf{p})$, then

$$\begin{aligned} \dot{\mathbf{W}} &= D\mathbf{f}\,(\Phi_\xi(\cdot,\mathbf{p}),\mathbf{p})\,\mathbf{W}\ . \\ \Rightarrow\quad sp\left(\dot{\mathbf{W}}\,\mathbf{W}^{-1}\right) &= sp\,(D\mathbf{f}\,(\Phi_\xi(\cdot,\mathbf{p}),\mathbf{p}))\ , \\ &= div\,\mathbf{f}\,(\Phi_\xi(\cdot,\mathbf{p}),\mathbf{p})\ . \end{aligned}$$

If $\mathbf{T}(t) \in \mathbb{R}^{n,n}$ is the transformation matrix that transforms $\mathbf{W}(t)$ for each time t into diagonal form $\boldsymbol{\Lambda}(t) = \mathbf{T}(t)\,\mathbf{W}(t)\,\mathbf{T}^{-1}(t)$, then

$$sp\left(\dot{\mathbf{W}}\mathbf{W}^{-1}\right) = sp\left(\mathbf{T}\,\dot{\mathbf{W}}\,\mathbf{T}^{-1}\,\boldsymbol{\Lambda}^{-1}\right)\ .$$

Furthermore we have

$$
\begin{aligned}
(i) \qquad (\mathbf{TWT}^{-1})^{\cdot} &= \dot{\mathbf{T}}\mathbf{WT}^{-1} + \mathbf{T}\dot{\mathbf{W}}\mathbf{T}^{-1} + \mathbf{TW}\left(\mathbf{T}^{-1}\right)^{\cdot} \;, \\
&= \left(\dot{\mathbf{T}}\mathbf{T}^{-1}\right)\boldsymbol{\Lambda} + \mathbf{T}\dot{\mathbf{W}}\mathbf{T}^{-1} + \boldsymbol{\Lambda}\mathbf{T}\left(\mathbf{T}^{-1}\right)^{\cdot} \;, \\
&= \left(\dot{\mathbf{T}}\mathbf{T}^{-1}\right)\boldsymbol{\Lambda} - \boldsymbol{\Lambda}\left(\dot{\mathbf{T}}\mathbf{T}^{-1}\right) + \mathbf{T}\dot{\mathbf{W}}\mathbf{T}^{-1} \;, \\
(ii) \qquad (\mathbf{TWT}^{-1})^{\cdot} &= \dot{\boldsymbol{\Lambda}} \;. \\
\Rightarrow \qquad \mathbf{T}\dot{\mathbf{W}}\mathbf{T}^{-1}\boldsymbol{\Lambda}^{-1} &= \dot{\boldsymbol{\Lambda}}\boldsymbol{\Lambda}^{-1} + \boldsymbol{\Lambda}\left(\dot{\mathbf{T}}\mathbf{T}^{-1}\right)\boldsymbol{\Lambda}^{-1} - \dot{\mathbf{T}}\mathbf{T}^{-1} \;. \\
\Rightarrow \qquad sp\left(\mathbf{T}\dot{\mathbf{W}}\mathbf{T}^{-1}\boldsymbol{\Lambda}^{-1}\right) &= sp\dot{\boldsymbol{\Lambda}}\boldsymbol{\Lambda}^{-1} + \underbrace{sp\left(\boldsymbol{\Lambda}\dot{\mathbf{T}}\mathbf{T}^{-1}\boldsymbol{\Lambda}^{-1}\right)}_{sp\left(\dot{\mathbf{T}}\mathbf{T}^{-1}\right)} - sp\left(\dot{\mathbf{T}}\mathbf{T}^{-1}\right) \;, \\
&= sp\left(\dot{\boldsymbol{\Lambda}}\boldsymbol{\Lambda}^{-1}\right) \;. \\
\Rightarrow \qquad div\,\mathbf{f} &= sp\left(\dot{\boldsymbol{\Lambda}}\boldsymbol{\Lambda}^{-1}\right) \;, \\
&= \sum_{i=1}^{2f} \frac{\dot{\lambda}_i}{\lambda_i} \;. \\
\Rightarrow \qquad \int_0^T div\,\mathbf{f}\,dt &= \sum_{i=1}^{2f} \int_{\lambda_i(0)}^{\lambda_i(T)} \frac{d\lambda_i}{\lambda_i} \;, \\
&= \sum_{i=1}^{2f} ln\left(\frac{\lambda_i(T)}{\lambda_i(0)}\right) \;, \\
&= ln\left(\prod_{i=1}^{2f}\lambda_i(T)\right) \quad (\lambda_i(0) = 1 \text{ or } \mathbf{W}(0) = \mathbf{E}!) \\
\Rightarrow \qquad e^{\int_0^T div\mathbf{f}\,dt} &= \prod_{i=1}^{2f}\lambda_i(T) \;, \\
&= exp\left(\sum_{i=1}^{2f}\mu_i\right) \;. \\
\Rightarrow \qquad \int_o^T div\,\mathbf{f}\,dt &= \sum_{i=1}^{2f}\mu_i \;.
\end{aligned}
$$

□

To minimize the functional $F(\mathbf{x},\mathbf{p}) = \int_0^T div\mathbf{f}(\mathbf{x},\mathbf{p})dt$ means to minimize the FLOQUET-exponents with respect to their arithmetic mean value. Hence, if the limit cycle is stable, its stability is maximal with respect to the applied criterion.

The problem now is to find a periodic solution $\Phi_\xi(\cdot,\mathbf{p})$ with time of period $T_\mathbf{p}$ and a parameter $\mathbf{p}^* \in P$ such that:

$$F\left(\Phi_\xi(\cdot,\mathbf{p}^*),\mathbf{p}^*\right) = opt\left\{F\left(\Phi_\xi(\cdot,\mathbf{p}),\mathbf{p}\right) | \left(\Phi_\xi(\cdot,\mathbf{p}),\mathbf{p}\right) \in C^1(\mathbb{R},TM) \times P\right\} \quad (2.119)$$

if subjected to

$$\frac{d}{dt}\Phi_\xi(t,\mathbf{p}^*) = \mathbf{f}\left(\Phi_\xi(t,\mathbf{p}^*),\mathbf{p}^*\right) \ . \quad (2.120)$$

"*opt*" represents a condition for the functional F (for instance *opt* = extremum).

2.7.2 Formulation of the problem as a boundary value problem

The computation of periodic solutions with additional properties according to chapter 2.7.1 turns out to be a standard problem of optimization. If a vector $\boldsymbol{\lambda} \in C^1(\mathbb{R},TM)$ of adjoint variables and the HAMILTONian function

$$\mathrm{H}(\mathbf{x},\boldsymbol{\lambda},\mathbf{p}) := \mathrm{g}(\mathbf{x},\mathbf{p}) - \boldsymbol{\lambda}^T\mathbf{f}(\mathbf{x},\mathbf{p}) \quad (2.121)$$

is introduced, the problem can be expressed by a boundary value problem as follows:

ODE	$\dot{\mathbf{z}}$	=	$\mathbf{J}\,D_\mathbf{z}\mathrm{H}(\mathbf{z},\mathbf{p})$
AC	0	=	$D_\mathbf{p}\mathrm{H}(\mathbf{z},\mathbf{p})$
BC	$\mathbf{x}(T)$ $\boldsymbol{\lambda}(T)$	= =	$\mathbf{x}(0) + 2\mathbf{k}\pi$ $\boldsymbol{\lambda}(0)$
TC	0	=	$\mathrm{H}(\mathbf{z}(\mathbf{T}),\mathbf{p})$

(2.122)

The components of the vector $\mathbf{z}$ are the state vector $\mathbf{x}$ and the adjoint variables $\boldsymbol{\lambda}$. $\mathbf{J}$ is the symplectic matrix

$$\mathbf{J} := \begin{bmatrix} \mathbf{0} & -\mathbf{E}_{2f} \\ \mathbf{E}_{2f} & \mathbf{0} \end{bmatrix} \in \mathbb{R}^{4f,4f} \ . \quad (2.123)$$

The augmented dynamical system is represented by equation ODE in equation (2.122). The HAMILTONian function H is constant with respect to $\mathbf{p}$. Hence, for each time t the vector $\mathbf{p}^*$ can be computed by means of the algebraic constraint (AC).

An interesting aspect is that the condition of periodicity for the state variables is also the case for the adjoint variables (BC). Finally, the transversality condition (TC) is necessary to compute the time of period T. The transversality condition holds for each time, i.e. H is constant along the trajectory. Or, in other words: The flow of the dynamical system takes place on the surface $\mathrm{H}^{-1}(0)$.

Therefore, the computation of periodic solutions of an arbitrary (!) differentiable non-linear dynamical system with special properties leads to the computation of a periodic solution of a HAMILTONian system on the surface $\mathrm{H}^{-1}(0)$. This problem has been studied in chapter 2.5. The method used there can be applied directly and without any restriction to solve this problem.

2.7.3 Singularities of the HAMILTONian system of optimization

The regularity of the symplectic matrix $\mathbf{J}$ leads to a singular point of the HAMILTONian system (2.122)

$$\mathbf{z}_0 = \begin{bmatrix} \mathbf{x}_0 \\ \boldsymbol{\lambda}_0 \end{bmatrix} \quad \in \quad \mathbb{R}^{4f} \tag{2.124}$$

if the gradient of the HAMILTONian function H vanishes at $\mathbf{z}_0$. That is (cf. equation (2.121)):

$$D_{\mathbf{z}}\mathrm{H}\,(\mathbf{z}_0) = \mathbf{0} \iff \begin{cases} (i) & \mathbf{f}(\mathbf{x}_0) = \mathbf{0}\ , \\ (ii) & \boldsymbol{\lambda}_0^T D_{\mathbf{x}}\mathbf{f}(\mathbf{x}_0) = D_{\mathbf{x}}\mathbf{g}(\mathbf{x}_0)\ . \end{cases} \tag{2.125}$$

Hence, if $\mathbf{z}_0$ is a singularity of the system described by equation (2.121), then $\mathbf{x}_0$ is a singularity of the vecorfield $\mathbf{f}$:

Because of

$$D^2_{\mathbf{z}}\mathrm{H}(\mathbf{z}_0) = \left[\begin{array}{c|c} D_{\mathbf{x}}\mathbf{f}(\mathbf{x}_0) & 0 \\ \hline D^2_{\mathbf{x}}g(\mathbf{x}_0) - D_{\mathbf{x}}\left(\boldsymbol{\lambda}_0^T D_{\mathbf{x}}\mathbf{f}(\mathbf{x}_0)\right) & -D_{\mathbf{x}}\mathbf{f}(\mathbf{x}_0) \end{array}\right] \tag{2.126}$$

the HESSEian matrix is regular iff $D_{\mathbf{x}}\mathbf{f}(\mathbf{x}_0)$ is regular. Therefore:

$$\mathbf{z}_0 \text{ is not degenerated} \iff \mathbf{x}_0 \text{ is not degenerated}\ .$$

The eigenvalues of $D^2_{\mathbf{z}}\mathrm{H}(\mathbf{z}_0)$ are determined by the eigenvalues of $D_{\mathbf{x}}\mathbf{f}(\mathbf{x}_0)$. Hence, (cf. equation (2.66)):

$$D_{\mathbf{z}}\mathrm{H}(\mathbf{z}) \text{ is resonant } \iff \mathbf{f}(\mathbf{x}) \text{ is resonant } .$$

According to the theory of normal forms the HAMILTONian system of optimization is diffeomorph to a linear system iff the non-linear dynamical system is diffeomorph to its linear part.

2.7.4 Examples of application

∇

Example 2.19: "conservative system"

Among all periodic solutions of the double pendulum according to chapter 2.5.3, case 2 a periodic solution with minimal time of period exists. In order to compute this periodic solution, the system is transformed into surface oriented coordinates of the energy surface $H^{-1}(h)$

$$\dot{\boldsymbol{\eta}} = \hat{\mathbf{f}}(\boldsymbol{\eta}, h) \ . \tag{2.127}$$

Then the HAMILTONian function of optimization is

$$\mathrm{H}(\boldsymbol{\eta}, \boldsymbol{\lambda}, \mathrm{h}) = 1 - \boldsymbol{\lambda}^{\mathrm{T}}\hat{\mathbf{f}}(\boldsymbol{\eta}, \mathrm{h}) \ , \tag{2.128}$$

$$\left(\boldsymbol{\eta}, \boldsymbol{\lambda} \in C^1(\mathbb{R}, \mathbb{R}^3)\right) \ .$$

The parameter $\mathbf{p}$ corresponds to the energy value h, obtained by

$$D_h\mathrm{H}(\boldsymbol{\eta}(\mathrm{t}),\ \boldsymbol{\lambda}(\mathrm{t}),\ \mathrm{h}) = 0 \tag{2.129}$$

($t \in [0, T]$ arbitrary).

ODE and boundary values are given by:

$$\begin{bmatrix} \dot{\boldsymbol{\eta}} \\ \dot{\boldsymbol{\lambda}} \end{bmatrix} = \begin{bmatrix} \hat{\mathbf{f}}(\boldsymbol{\eta}, h) \\ -\left[D_{\boldsymbol{\eta}}\hat{f}(\boldsymbol{\eta}, h)\right]^T \boldsymbol{\lambda} \end{bmatrix} , \tag{2.130}$$

$$\mathbf{r}\left(\boldsymbol{\eta}(0),\ \boldsymbol{\eta}(T),\ \boldsymbol{\lambda}(0),\ \boldsymbol{\lambda}(T)\right) = \begin{bmatrix} \boldsymbol{\eta}(T) - \boldsymbol{\eta}(0) - 2\mathbf{k}\pi \\ \boldsymbol{\lambda}(T) - \boldsymbol{\lambda}(0) \\ \eta_3(0) - \frac{2k_1\pi}{T} \end{bmatrix} = \mathbf{0} \ . \tag{2.131}$$

The periodic solution is incorporated in the surface $H^{-1}(0)$. Therefore it is useful to make an additional transformation into surface oriented coordinates (cf. equation (2.98)). However, for the purpose of numerical computation it is more convenient to plug the parameter h into the boundary value condition, since $H(\boldsymbol{\eta}(t), \boldsymbol{\lambda}(t), h) = const.$ with respect to h. And the transversality condition says that this constant value must be zero. Hence, the transversality condition (equation (2.122)) can be used to compute h.

h is not a function of t, therefore $\dot{h} = 0$ is an additional ODE. And the corresponding boundary value condition is the transversality condition.

If the time interval $[0, T]$ is transformed into the unit time interval $[0, 1]$ the complete BVP (cf. chapter 2.5.3) is given by:

$$\begin{bmatrix} \boldsymbol{\eta}' \\ \boldsymbol{\lambda}' \\ T' \\ h' \end{bmatrix} = \begin{bmatrix} \hat{\mathbf{f}}(\boldsymbol{\eta}, h)\, T \\ -\left[D_{\boldsymbol{\eta}}\hat{\mathbf{f}}(\boldsymbol{\eta}, h)\right]^T \boldsymbol{\lambda}\, T \\ 0 \\ 0 \end{bmatrix} \quad \left(' := \frac{d}{d\tau}\right) , \tag{2.132}$$

$$\mathbf{r}\,(\boldsymbol{\eta}(0),\, \boldsymbol{\eta}(1),\, \boldsymbol{\lambda}(0),\, \boldsymbol{\lambda}(1)) = \begin{bmatrix} \boldsymbol{\eta}(1) - \boldsymbol{\eta}(0) - 2\mathbf{k}\pi \\ \boldsymbol{\lambda}(1) - \boldsymbol{\lambda}(0) \\ \eta_3(0) - 2k_1\pi \\ H(\boldsymbol{\eta}(0),\, \boldsymbol{\lambda}(0),\, h(0)) \end{bmatrix} = \mathbf{0}\ . \tag{2.133}$$

The BVP for computation of bifurcation points investigated by [SEYDEL 1979a] and the one shown here are exactly the same provided that the time of period T and/or the energy value h is used as the bifurcation parameter. This result is not restricted to this example. That means the computation of periodic solutions is equivalent to the computation of bifurcation points or more precise to the computation of turning points. And in HAMILTONian systems the time of period represents the bifurcation parameter.

After bifurcation a qualitative change of the phase portrait will take place. This can be visualized for instance by projection of the phase curve of the double pendulum into the $(q_2, \dot{q}_2)$–plane (fig. 2.30).

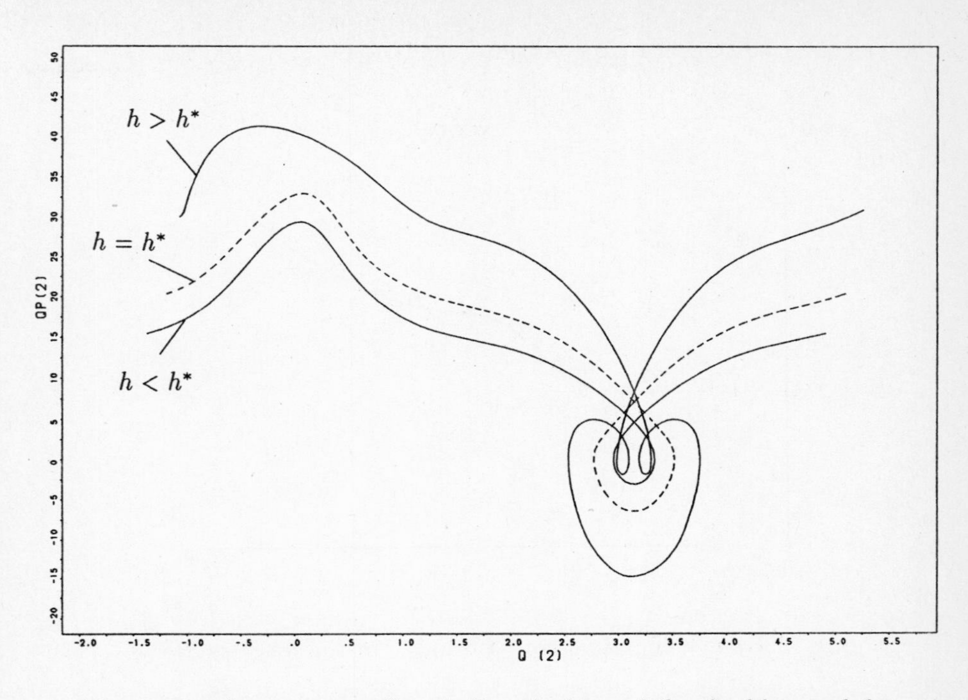

Fig. 2.30: phase curves of periodic solutions of the double pendulum
$(h < h^*,\ h = h^*,\ \ h > h^*)$

At the upper turning point $q_2 = \pi$ of the second pendulum the behavior of motion on the energy surface $H^{-1}(h = h^*)$ changes from one libration to three librations.

$\triangle$

∇

Example 2.20: (dissipative and excited systems)

Corresponding to example 2.19, a limit cycle with minimal time of period of the wheel set discussed in chapter 2.6.3 can be computed.

At velocity $v_0 = v_0^* = 19\frac{m}{s}$ a bifurcation occurs from the stable straight forward run to a superposed periodic lateral deviation. The time of period T immediately after bifurcation is $0.78s$. That the limit cycles after the bifurcation are unstable (broken line in fig. 2.31).

The limit cycles become stable at a velocity $v_0 \approx 21\frac{m}{s}$. At this point the time of period T is minimal: $T_{min} \approx 0.74s$. Actually the bifurcation point is a turning point. A second turning point takes place at $v_0 \approx 23\frac{m}{s}$. At this velocity the lateral deviation x or γ respectively an additional cycle at each side (cf. fig. 2.31) occurs.

The unstable periodic solutions are the reason for amplitude jumps that are well known in the theory of non-linear vibrations.

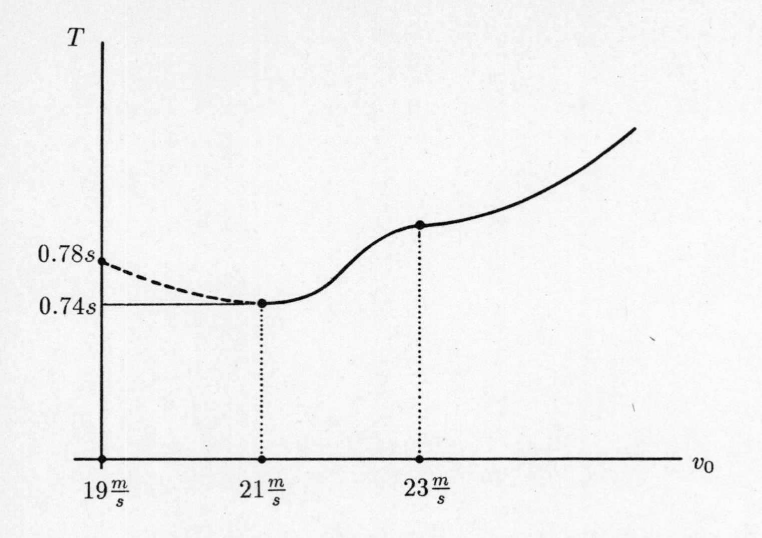

Fig. 2.31: velocity v_0 against period T of the limit cycles

△

Namely, each solution of an initial value problem converges to a stable limit cycle, never against an unstable limit cycle. For $v_0 < 21\ \frac{m}{s}$ only the fix point is stable and the limit cycle is unstable. That is each solution will be attracted by the fix point. For $v_0 > 21\ \frac{m}{s}$ the limit cycle becomes stable and now each solution will be attracted by the cycle. That means a jump in the amplitude will occur because the limit cycle has a non zero amplitude immediately after its stabilization.

2.8 Stability and bifurcation of periodic solutions

A solution $\Phi_\xi : [0,\infty[\rightarrow TM$ of the dynamical system (2.27) with initial point $\boldsymbol{\xi} \in TM$ $(\Phi_\xi(0) = \boldsymbol{\xi})$ is called "asymptotically stable" iff:

(a) $\{\Phi_\xi(t) | t > 0\}$ is bounded (with respect to a norm $\| \cdot \|$ on TM) .

(b) There is a $\varepsilon > 0$ such that for each solution Φ_ξ with $\| \boldsymbol{\xi} - \boldsymbol{\xi}' \| < \varepsilon$

$$\lim_{t\to\infty} \| \Phi_\xi(t) - \Phi_{\xi'}(t^*) \| = 0$$

is the case.

The time function $t^* : \mathbb{R}_+ \to \mathbb{R}_+$ is defined by

$$\| \Phi_\xi(t) - \Phi_{\xi'}(t^*) \| := \min\left\{ \| \Phi_\xi(t) - \Phi_{\xi'}(t') \| \;\middle|\; t' > 0 \right\}$$

(cf. fig. 2.32).

The definition given here is equivalent to the standard definition given in literature. For $t = t^*$ the system is called LIAPUNOV–stable.

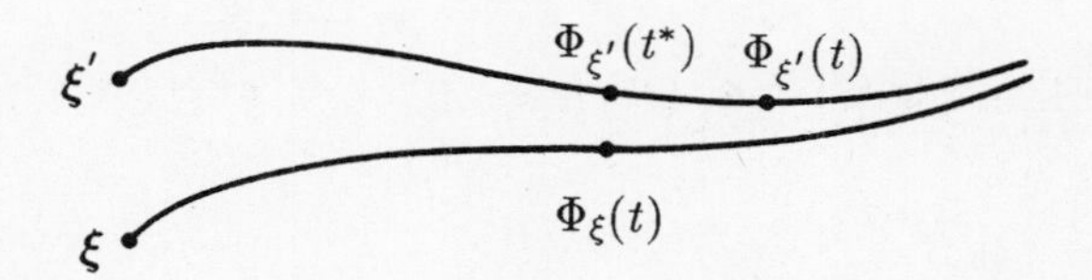

Fig. 2.32: definition of asymptotic stability

The value ε is a measure of the stability of Φ_ξ. The ball $K(\boldsymbol{\xi}) := \left\{\boldsymbol{\xi} \in TM | \, \| \boldsymbol{\xi} - \boldsymbol{\xi}' \| < \varepsilon\right\}$ represents a domain of attraction of a stable solution Φ_ξ. $K(\boldsymbol{\xi})$ is an open set, therefore $\Phi_{\xi'}$ is also a stable solution for all $\boldsymbol{\xi}' \in K(\boldsymbol{\xi})$, probably with another measure ε' of stability. The set of all connected domains of attraction $K(\boldsymbol{\xi})$ (=: connected component Z^+) is called the "stability domain". It is not necessary that all domains of attraction are connected. Therefore more of them may exist. The stability definition given above defines a connection between the stability domain and the fixed points or periodic solutions of the system. A more precise formulation is stated in the following theorem.

Theorem (2.4)

If the solutions of the dynamical system $\dot{\mathbf{x}} = \mathbf{f}(\mathbf{x}, \mathbf{p})$ are unique for any arbitrary initial point $\boldsymbol{\xi} \in Z^+$ then the following is true:

(1) Inside each stability domain Z^+ mentioned above a limit function Φ_ξ exists such that each solution $\Phi_{\xi'}$ with $\boldsymbol{\xi}' \in Z^+$ converges Φ_ξ for $t \to \infty$.

(2) Either the limit function Φ_ξ is a fixed point or a periodic solution (equilibrium between dissipated and supplied energy). Φ_ξ is unique.

Proof:

to (1):

Assume that $(t_\nu)_{\nu \in \mathbb{N}}$ is an arbitrary series of points of time $t_\nu > 0$ so that $t_{\nu+1} > t_\nu \quad \forall\ \nu \in \mathbb{N}$.

If $\Phi_{\mathbf{z}}$ is the solution that belongs to an arbitrary initial point $\mathbf{z} \in Z^+$, then

$$\mathbf{z}_\nu := \Phi_{\mathbf{z}}(t_\nu) \ ; \quad (\nu = 0, 1, 2, \ldots) \tag{2.134}$$

is the series of the points in the state space which corresponds to the time series.

Sequence $(\mathbf{z}_\nu)_{\nu \in \mathbb{N}}$ is bounded because $\Phi_{\mathbf{z}}$ is bounded according to definition (a). Hence, there is at least one cluster point $\boldsymbol{\xi} \in Z^+$ and a partial sequence – for instance $\left(\mathbf{z}_{\sigma(k)}\right)_{k \in \mathbb{N}}$, where $\sigma : \mathbb{N} \to \mathbb{N}$ is an injective and a monotonously increasing index function – that converges to $\boldsymbol{\xi}$.

Φ_ξ is this limit function since, according to definition (b):

$$\lim_{t \to \infty} \| \Phi_\xi(t) - \Phi_{\xi'}(t^*) \| = 0 \qquad \forall\ \boldsymbol{\xi}' \in Z^+ \ .$$

to (2):

There are two possibilities for the limit function.

1st possibility:

The cluster point $\boldsymbol{\xi}$ is a fixed point $\Rightarrow \Phi_\xi(t) = \boldsymbol{\xi} \ \ \forall\, t > 0$. $\Rightarrow$ limit function Φ_ξ is the fixedd point $\boldsymbol{\xi}$ and the system is LIAPUNOV–stable in Z^+.

2nd possibility:

The cluster point is not a fixed point.

In that case we consider a manifold Ω of codimension 1, that intersects the trajectory Φ_ξ at $\boldsymbol{\xi}$ transversally (cf. fig. 2.33).

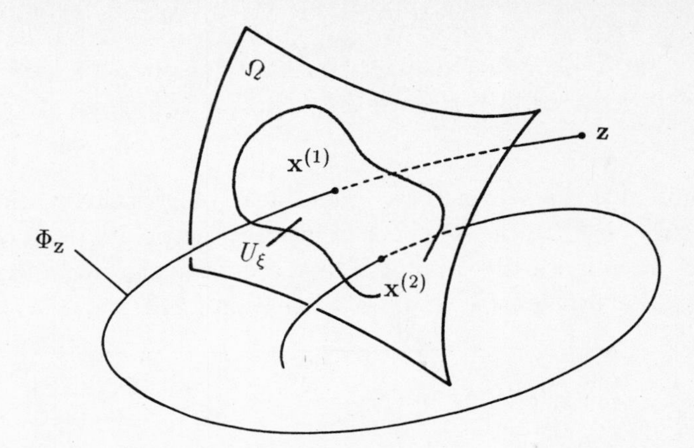

Fig. 2.33: intersecting points $\mathbf{x}_1, \mathbf{x}_2, \ldots,$ through Ω

Then an open neighborhood U_ξ of $\boldsymbol{\xi}$ with $U_\xi \subset \Omega \cap Z^+$ exists which

1. does not contain another cluster point and
2. is intersected transversally by any trajectory $\Phi_{\xi'}$ with $\boldsymbol{\xi}' \in Z^+$ at a certain point of time $t(\boldsymbol{\xi}')$.

$\boldsymbol{\xi}$ is the limit of a special sequence defined by $(\mathbf{x}_\nu)_{\nu\in\mathbb{N}}$ where each $\mathbf{x}_\nu$ is a intersecting point of the function $\Phi_\mathbf{z}$ ($\mathbf{z} \in Z^+$ arbitrary) and $\mathbf{x}_\nu \in U_\xi$ for all $\nu \in \mathbb{N}$. Namely if $(t_\nu)_{\nu\in\mathbb{N}}$ is the time sequence generated by $(\mathbf{x}_\nu)_{\nu\in\mathbb{N}}$ ($\mathbf{x}_\nu = \Phi_\mathbf{z}(t_\nu)$), definition (b) leads to:

$$\begin{aligned} 0 &= \lim_{\nu\to\infty} \| \Phi_\mathbf{z}(t_\nu) - \Phi_\xi(t_\nu) \| \ , \\ &= \lim_{\nu\to\infty} \| \mathbf{x}_\nu - \Phi_\xi(t_\nu) \| \ , \\ &= \| \lim_{\nu\to\infty} \mathbf{x}_\nu - \boldsymbol{\xi} \| \ . \end{aligned}$$

$$\Rightarrow \quad \lim_{\nu\to\infty} \mathbf{x}_\nu = \boldsymbol{\xi} \ . \tag{2.135}$$

If we choose $\mathbf{z} := \boldsymbol{\xi}$, then $\boldsymbol{\xi}$ is fixed point, that is $\Phi_\xi(t_\nu) = \boldsymbol{\xi} \ \ \forall \ \nu \in \mathbb{N}$ (remember that there is no other cluster point in U_ξ). That can be only the case if Φ_ξ is a periodic solution.

If $(t_\nu)_{\nu\in\mathbb{N}}$ is the time sequence generated by $\Phi_\mathbf{z}$ ($\mathbf{z} \in Z^+$ arbitrary) and the intersecting points $\mathbf{x}_\nu$, then the period T is given by

$$T = \lim_{\nu\to\infty} (t_{\nu+1} - t_\nu) \ . \tag{2.136}$$

□

Notes:

1. Areas Z^- that are stability domains if time t runs into "negative direction", that is if t is replaced by $-t$, have the same behavior than stability domains Z^+.

2. If the manifold Ω is a zero set of a – non-linear – algebraic function $g : TM \to \mathbb{R}$, that is $\Omega = g^{-1}(0)$, the sequence of intersecting points $(\mathbf{x}_n)_{n\in\mathbb{N}}$ with $\mathbf{x}_n := \Phi_{\mathbf{z}}(t_n)$ can be numerically computed by a recursive algorithm that consists of boundary value problems according to equation (2.137):

$$\boxed{\begin{array}{l} \mathbf{x}_0 := \boldsymbol{\xi} \\ \underline{i = 1, 2, \ldots} \\ \qquad \begin{bmatrix} \mathbf{x} \\ t_i \end{bmatrix}' = \begin{bmatrix} t_i \mathbf{f}(\mathbf{x}, \mathbf{p}) \\ 0 \end{bmatrix} \\ \qquad \mathbf{r}\,(\mathbf{x}(0), \mathbf{x}(1)) := \begin{bmatrix} \mathbf{x}(0) - \mathbf{x}_{i-1} \\ g(\mathbf{x}(1)) \end{bmatrix} = \mathbf{0} \\ \qquad \mathbf{x}_i := \mathbf{x}(1) \end{array}} \tag{2.137}$$

The algorithm computes the sequence $(\mathbf{x}_i)_{i\in\mathbb{N}}$ as well as the time sequence $(t_i)_{i\in\mathbb{N}}$.

3. The easiest choice for Ω is a linear space that is perpendicularly crossed by Φ_ξ at $\boldsymbol{\xi}$. In that case Ω is called a POINCARÉ–section and g is defined by

$$g(\mathbf{x}) := (\mathbf{x} - \boldsymbol{\xi})^T \mathbf{f}(\boldsymbol{\xi}, \mathbf{p}) \ .$$

4. If the recursion, defined by the algorithm according to equation (2.137), is represented by a map

$$P : \Omega \to \Omega \tag{2.138}$$

and $d_\Omega : \Omega \to \mathbb{R}$ is a metric on Ω, then the stability definition (a) and (b) on $Z^+ \cap \Omega$, mentioned above, is equivalent to:

A sequence $(\mathbf{x}_n)_{n\in\mathbb{N}} := \left(P^n(\boldsymbol{\xi})\right)_{n\in\mathbb{N}}$ is called asymptotically stable if

(a') the sequence $(\mathbf{x}_n)_{n\in\mathbb{N}}$ is bounded.

(b') There is an $\varepsilon > 0$ so that for each neighboring sequence $\left(\mathbf{x}_n^{'}\right)_{n\in\mathbb{N}} := \left(P^n(\boldsymbol{\xi}^{'})\right)_{n\in\mathbb{N}}$ with $\| \boldsymbol{\xi} - \boldsymbol{\xi}^{'} \| < \varepsilon$

$$\lim_{n\to\infty} d_\Omega \left(P^n(\boldsymbol{\xi}), P^n(\boldsymbol{\xi}^{'})\right) = 0 \ .$$

It is advantageous to use this definition if differentiable systems are investigated that have discontinuities with respect to the state space variables $\mathbf{x}$ (cf. chapter 3).

5. According to theorem 2.4 the dynamical behavior inside stability domains is described or "governed" in an unique way by the fixed points and periodic solutions of the system. Namely, the theory of normal forms of POINCARÉ and SIEGEL (cf. chapter 2.3 and chapter 2.8.3) proofs the equivalence between the non-linear system and the linearized system around the fixed point or around the periodic solution

   A limit function (fixed point or periodic solution) inside a stable domain is of course also stable. If the limit function loses stability – for instance by variation of the system parameter $\mathbf{p}$ – in at least one direction of the state space TM, the domain of stability will be reduced in its dimension (stable and unstable manifolds). In other words: number and distribution of stable fixed points and periodic solutions as well as the stability domains belonging to them define a "measure of regularity" of all solutions of the dynamical system.

2.8.1 Monodromy matrix and stability of a periodic solution

Assume that Φ_ξ is a periodic solution of equation (2.27) with period T. $\boldsymbol{\xi}$ is a given but arbitrary initial point on the periodic solution. The linear space

$$E := \left\{\boldsymbol{\xi} + \mathbf{y} \in \mathbb{R}^{2f} \middle| \mathbf{y}^T\mathbf{f}(\boldsymbol{\xi}, \mathbf{p}) = \mathbf{0}\right\} \ , \tag{2.139}$$

that is vertically intersected by Φ_ξ at the point $\boldsymbol{\xi}$, is mapped by the phase flow Φ at time T into a manifold Ω. In that case, Ω is defined by

$$\Omega := \{\Phi_{\xi+y}(T) | \boldsymbol{\xi} + \mathbf{y} \in E\} \ . \tag{2.140}$$

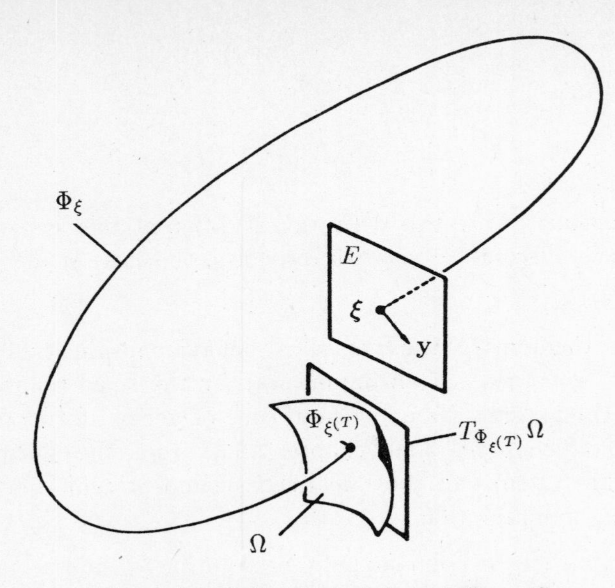

Fig. 2.34: definition of Ω and $T_{\Phi_\xi(T)}\Omega$

Regarding the tangent space $T_{\Phi_\xi(T)}\Omega$ at $\Phi_\xi(T) = \boldsymbol{\xi} + 2\mathbf{k}\pi$ of Ω the following theorem is of interest:

theorem (2.5):

If $(\mathbf{u}_1, \ldots, \mathbf{u}_{2f-1})$ is a frame of E, then $(\mathbf{v}_1, \ldots, \mathbf{v}_{2f-1})$ is a frame of $T_{\Phi_\xi(T)}\Omega$. $\mathbf{v}_i$ is defined by

$$\mathbf{v}_i := \mathbf{W}(T)\,\mathbf{u}_i \; ; \quad (i = 1, \ldots, 2f-1) \; . \tag{2.141}$$

$\mathbf{W}(T)$ is the WRONSKIan matrix of the linear time-variant ODE

$$\dot{\mathbf{z}}(t) = D\mathbf{f}\,(\Phi_\xi(t), \mathbf{p})\; \mathbf{z}(t) \tag{2.142}$$

at time T.

Proof:

If $[-s_0, s_0] \to E, s \mapsto \mathbf{x}(s)$; $s_0 > 0$ is an arbitrary piece of a curve intersecting the point $\boldsymbol{\xi}$ with $\mathbf{x}(0) = \boldsymbol{\xi}$, then, considering the curve $s \mapsto \Phi_{\mathbf{x}(s)}(T)$ on the manifold Ω we have

$$\dot{\Phi}_{\mathbf{x}(s)}(t) = \mathbf{f}\left(\Phi_{\mathbf{x}(s)}(t), \mathbf{p}\right) \; . \tag{2.143}$$

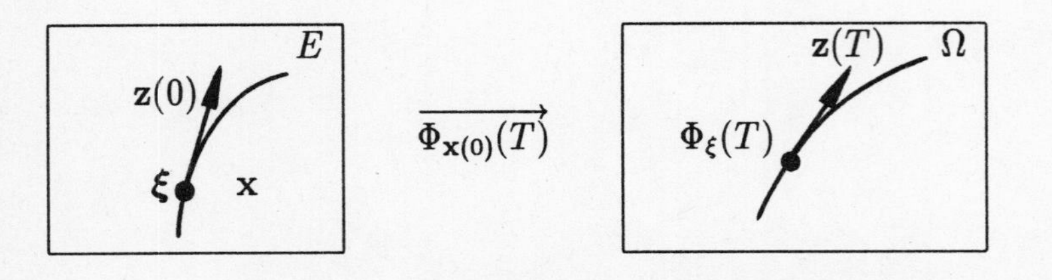

$$\Rightarrow \quad \frac{d}{dt}\left[\frac{d}{ds}\Phi_{\mathbf{x}(s)}(t)\bigg|_{s=0}\right] = D\mathbf{f}\left(\Phi_\xi(t),\mathbf{p}\right)\underbrace{\left[\frac{d}{ds}\Phi_{\mathbf{x}(s)}(t)\bigg|_{s=0}\right]}_{=:\mathbf{z}(t)} \tag{2.144}$$

$$\Rightarrow \quad \mathbf{z}(t) = \mathbf{W}(t)\cdot\mathbf{z}(0) \ . \tag{2.145}$$

$\mathbf{W}(t)$ is the WRONSKIan matrix of ODE (2.144).

The initial point $\mathbf{z}(0) = \frac{d}{ds}\Phi_{\mathbf{x}(s)}(0) = \frac{d\mathbf{x}}{ds}(0)$ is tangent vector at $\boldsymbol{\xi}$ of the curve $s \mapsto \mathbf{x}(s)$. $\mathbf{z}(T)$ is tangent vector at $\Phi_\xi(T)$ of the curve $s \mapsto \Phi_{\xi(s)}(T)$.

If $2f-1$ curves intersect point $\boldsymbol{\xi}$ of E and their tangent vectors $\mathbf{u}_i$ are linear independent (this is possible because of $dim E = 2f-1$), that is $(\mathbf{u}_1,\ldots,\mathbf{u}_{2f-1})$ is a frame of E, the transformed tangent vectors $\mathbf{v}_i := \mathbf{W}(T)\mathbf{u}_i$ establish a frame of $T_{\Phi_\xi(T)}\Omega$ (because of the regularity of $\mathbf{W}(T)$!).

□

The time history of a (small) disturbance of the initial point $\mathbf{y}(0) \in E$ is described by the linearized ODE around the periodic solution Φ_ξ:

$$\dot{\mathbf{y}}(t) = D\mathbf{f}\left(\Phi_\xi(t),\mathbf{p}\right)\cdot\mathbf{y}(t) \ . \tag{2.146}$$

Solution of equation (2.146) is

$$\mathbf{y}(t) = \mathbf{W}(t)\,\mathbf{y}(0) \ . \tag{2.147}$$

Hence, the disturbance $\mathbf{y}$ at time T is given by:

$$\mathbf{y}(T) = \mathbf{W}(T)\,\mathbf{y}(0) \quad \in \ T_{\Phi_\xi(T)}\Omega \ . \tag{2.148}$$

And the stability behavior of the periodic solution Φ_ξ is determined by the mapping

$$\mathbf{W}(T) : E \longrightarrow T_{\Phi_\xi(T)}\Omega \ . \tag{2.149}$$

$\mathbf{W}(T)$ is called the "monodromy operator" or "monodromy matrix" respectively.

If $\mathbf{W}(T)$ is transformed into diagonal form by an appropriate transformation matrix $\mathbf{U} \in \mathbb{C}^{2f,2f}$, equation (2.147) can be expressed by

$$\mathbf{z}(T) = \boldsymbol{\Lambda}\,\mathbf{z}(0) \tag{2.150}$$

where $\mathbf{z}(T) := \mathbf{U}\,\mathbf{y}(T) \in \mathbb{C}^{2f}$; $\mathbf{z}(0) := \mathbf{U}\,\mathbf{y}(0) \in \mathbb{C}^{2f}$ and $\boldsymbol{\Lambda} := \mathbf{U}\,\mathbf{W}(T)\mathbf{U}^{-1} = diag\,\{\lambda_1,\ldots,\lambda_{2f}\} \in \mathbb{C}^{2f,2f}$. Each component of equation (2.150) has the form

$$z_i(T) = \lambda_i\, z_i(0) \ . \tag{2.151}$$

$$\Rightarrow \qquad |z_i(T)| = |\lambda_i| \cdot |z_i(0)| \ . \tag{2.152}$$

That is the disturbance $z_i(0) \in \mathbb{C}$ increases if $|\lambda_i| > 1$ and decreases if $|\lambda_i| < 1$.

Therefore, stable periodic solutions are determined by eigenvalues located inside the unit circle and unstable periodic solutions are determined by eigenvalues located inside and outside the unit circle.

$\mathbf{W}(t)$ can be computed by means of the initial value problem

$$\dot{\mathbf{W}}(t) = D\mathbf{f}\left(\Phi_\xi(t), \mathbf{p}\right) \mathbf{W}(t) \ ; \quad \mathbf{W}(0) = \mathbf{E} \ . \tag{2.153}$$

First, the periodic function Φ_ξ must be computed, for instance by using the multiple shooting method. In case of numerical computation Φ_ξ is known only on a small number n $(n < 50$ mostly) of points

$$(t_i \ , \ \Phi_\xi(t_i)) \ ; \quad (i = 0, 1, \ldots, m) \ . \tag{2.154}$$

Hence, a convenient way to compute the monodromy matrix $\mathbf{W}(T)$ numerically is given by a sequence of n initial value problems

$$\begin{array}{l} t_0 := 0 \\ \mathbf{W}^{(0)}(t_0) := \mathbf{E} \\ \underline{i = 1, \ldots, n} \\ \quad t_i := \left(\frac{i}{n}\right) T \\ \quad \text{for } t \in [t_{i-1} \ , \ t_i] \ \text{ solve the initial value problem} \\ \quad \dot{\mathbf{x}} = \mathbf{f}(\mathbf{x}, \mathbf{p}) \qquad ; \ \mathbf{x}(t_{i-1}) = \Phi_\xi \left(t_{i-1}\right) \\ \quad \dot{\mathbf{W}}^{(i)} = D\mathbf{f}(\mathbf{x}, \mathbf{p}) \, \mathbf{W}^{(i)} \ ; \ \mathbf{W}^{(i)} \left(t_{i-1}\right) := \mathbf{W}^{(i-1)} \left(t_{i-1}\right) \\ \mathbf{W}(T) := \mathbf{W}^{(n)}(t_n) \end{array} \tag{2.155}$$

Also, $\mathbf{W}(T)$ can be computed by the initial value problem (2.153) if, for example, the exact function Φ_ξ is replaced by a HERMITE-interpolation $\tilde{\Phi}_\xi$ with cubic splines mounted on the points used in equation (2.154).

If the numerical computation is very time-consuming, it may be more efficient to approximate the JACOBIan matrix $D\mathbf{f}(\mathbf{x},\mathbf{p})$ by a cubic spline through the points

$$(t_i\ ,\ \ D\mathbf{f}(\Phi_\xi(t_i)\ ,\ \mathbf{p}))\ \ ;\quad (i=0,1,\ldots,m)\ .$$

A third possibility to compute $\mathbf{W}(T)$ is by using interim results of the multiple shooting method (cf. [SEYDEL 1988]).

∇

Example 2.21: stability of a limit cycle of the wheel set of a railway vehicle.

The left hand side of fig. 2.35 shows the phase curve for the lateral deviation of the wheel set of a railway vehicle, considered in chapter 2.6.2. The corresponding eigenvalues of the monodromy matrix inside the unit circle K_1 of $\mathbb{C}$ at a velocity $v_0 = 30\frac{m}{s}$ are shown on the right hand side of fig. 2.35.

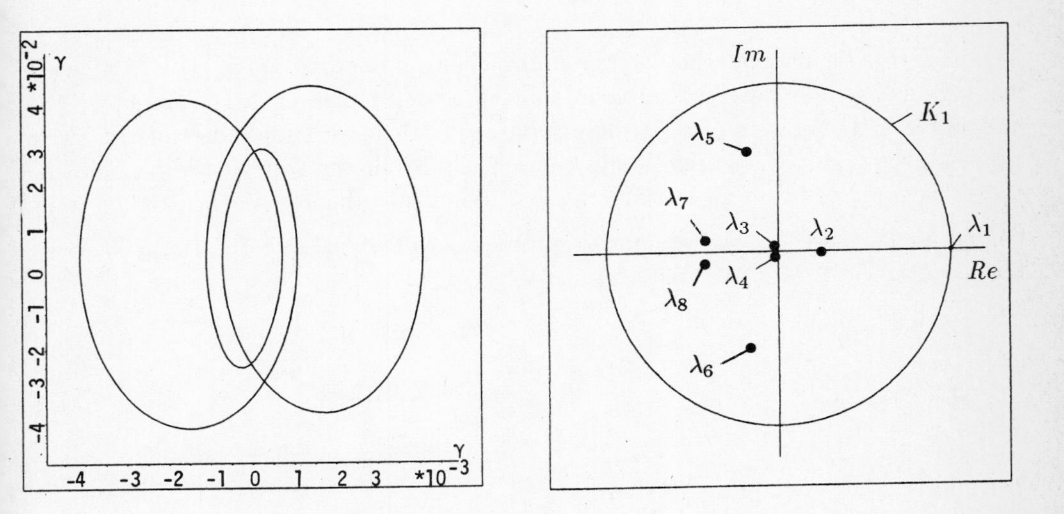

Fig. 2.35: phase curve of the lateral deviation x or γ respectively and eigenvalues of the corresponding monodromy matrix

Besides three conjugate complex pairs of eigenvalues (λ_3, $\lambda_4 = \overline{\lambda}_3$, λ_5, $\lambda_6 = \overline{\lambda}_5$, λ_7, $\lambda_8 = \overline{\lambda}_7$) and the expected 1–eigenvalue λ_1 there is an additional real eigenvalue λ_2. The modification of eigenvalue λ_2 by variation of v_0 leads to a "secondary" HOPF-bifurcation that is destabilization and splitting of the limit cycle (cf. chapter 2.8.2).

△

2.8.2 HOPF–bifurcation as consequence of loss of stability

A fix point $\mathbf{x}_0$ or a periodic solution Φ_ξ – if existent – (with $\Phi_\xi(0) = \boldsymbol{\xi} \in TM$) of the dynamical system (2.108) depends usually on a set of parameters $\mathbf{p}$ of the system. This is also the case for eigenvalues λ_i ($\mathbf{p} \mapsto \boldsymbol{\Lambda}(\mathbf{p})$) of the linearized system $\dot{\mathbf{y}} = D\mathbf{f}(\mathbf{x}_0)\mathbf{y}$ around $\mathbf{x}_0$ or of the monodromy matrix $\mathbf{W}(T)$. We assume that the parameter space is connected. Hence, if there are values $\mathbf{p} \in P$ that correspond to stable fix points or periodic solutions as well as values $\mathbf{p} \in P$ that correspond to unstable fix points and/or periodic solutions, curves $\boldsymbol{\gamma}$

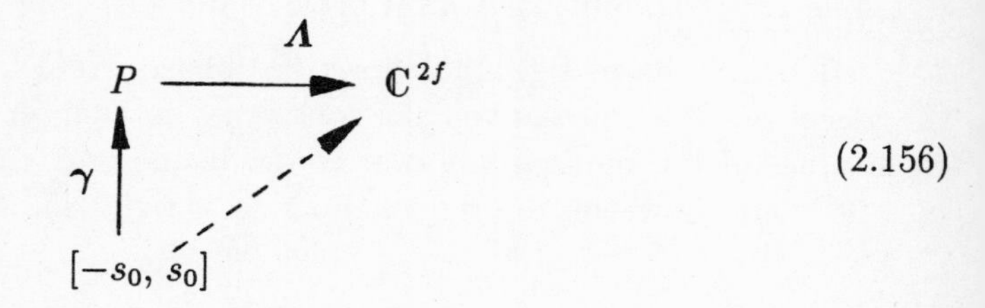

(2.156)

can be defined such that the corresponding eigenvalue curve $\lambda_i(\boldsymbol{\gamma}(s))$ – of at least one eigenvalue λ_i – intersects the imaginary axis (in case of a fix point) or the unit circle of $\mathbb{C}$ (in case of a periodic solution) with increasing values of the curve parameter s. The parametrization of curve γ can be chosen so that the fix point $\mathbf{x}_0$ or the periodic solution Φ_ξ is stable for $s < 0$ and unstable for $s > 0$.

For $Im\,\{\lambda_i(s)\} \neq 0$ we get two curves symmetric to the real axis (cf. fig. 2.36) since $D\mathbf{f}(\mathbf{x}_0)$ and $\mathbf{W}(T)$ are real matrices.

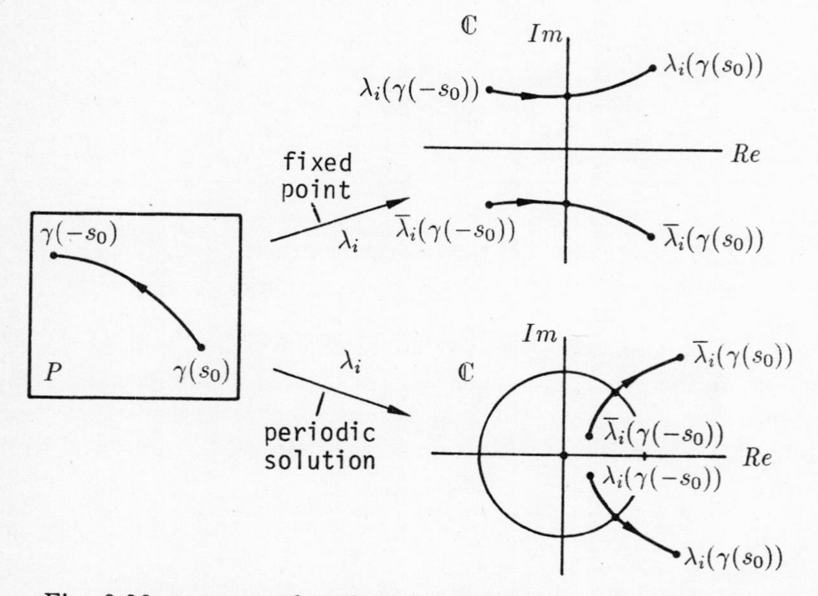

Fig. 2.36: transversal path of $\lambda_i(\boldsymbol{\gamma}(s))$ in case of a fix point and in case of a periodic solution

The described mechanism to destabilize a fixed point or a periodic solution of a non-linear dynamical system is very common, that is primary and secondary HOPF–bifurcations are a typical phenomenon in differentiable dynamical systems with dissipation and outside/inside excitation. The transversal intersection of eigenvalues λ_i at the imaginary axis with

$$\frac{d}{ds}\left(Re(\lambda_i \circ \gamma)\right)\bigg|_{s=0} = 0$$

in case of a fixed point or at the unit circle of $\mathbb{C}$ on the real axis, that is at

$$\begin{array}{ll} +1 & \text{(bifurcation, turning point)} \\ -1 & \text{(period doubling)} \end{array}$$

in case of a periodic solution happens mainly in HAMILTONian systems (cf. chapter 2.8.4) and in differentiable dynamical systems with discontinuities described by recursive functions (cf. chapter 3).

∇

Example 2.22: HOPF-bifurcation for a railway vehicle system

The stable straight-forward run of the wheel set depends greatly on the velocity of the railway vehicle. Therefore it is reasonable to use v_0 as bifurcation parameter.

(a) Primary HOPF-bifurcation

According to point (c) in chapter 2.6.3 $\mathbf{Y}_0$ is a fixed point of the vector field $\mathbf{F}$. The theorem of HOPF says that a bifurcation of a (stable) singular point $\mathbf{Y}_0$ to a periodic solution occurs iff at least one simple conjugate pair of eigenvalues $(\lambda, \overline{\lambda}) \in \mathbb{C}^2$ of $D\mathbf{F}(\mathbf{Y}_0, \mathbf{p})$ and a $v_0^* \in \mathbb{R}$ exists so that

$$(1) \qquad Re(\lambda(v_0)) \begin{cases} > 0 & \text{if} \quad v_0 > v_0^* \\ = 0 & \text{if} \quad v_0 = v_0^* \\ < 0 & \text{if} \quad v_0 < v_0^* \end{cases}$$

$$(2) \qquad \frac{d\,Re\lambda}{d\,v_0}(v_0^*) > 0 \ .$$

For the dataset, listed in the appendix, assumption (1) and (2) are true for the eigenvalue λ_1 at the value of the velocity $v_0^* = 19\frac{m}{s}$.

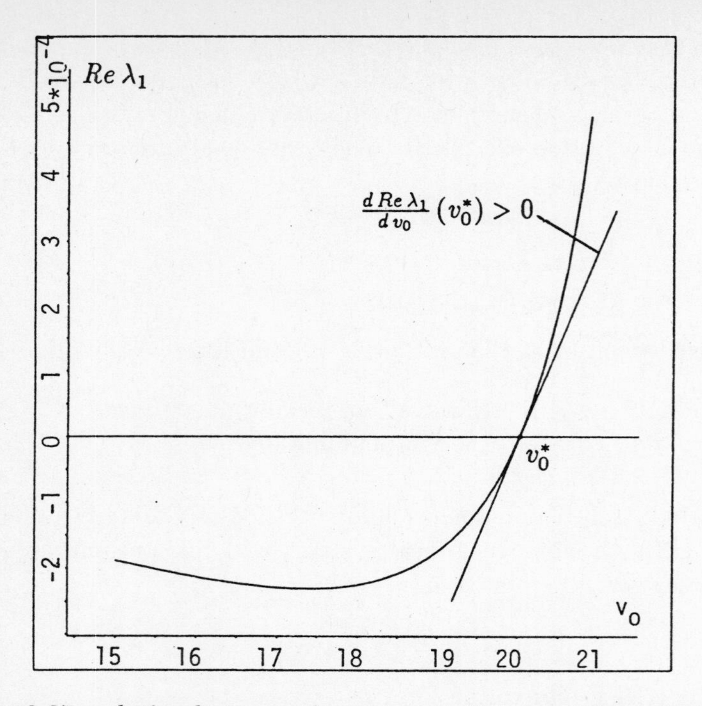

Fig. 2.37: relation between the real part $Re\lambda_1$ and the velocity v_0 of the railway vehicle

Fig. 2.38 shows the behavior of all eigenvalues in $\mathbb{C}$ with increasing velocity v_0.

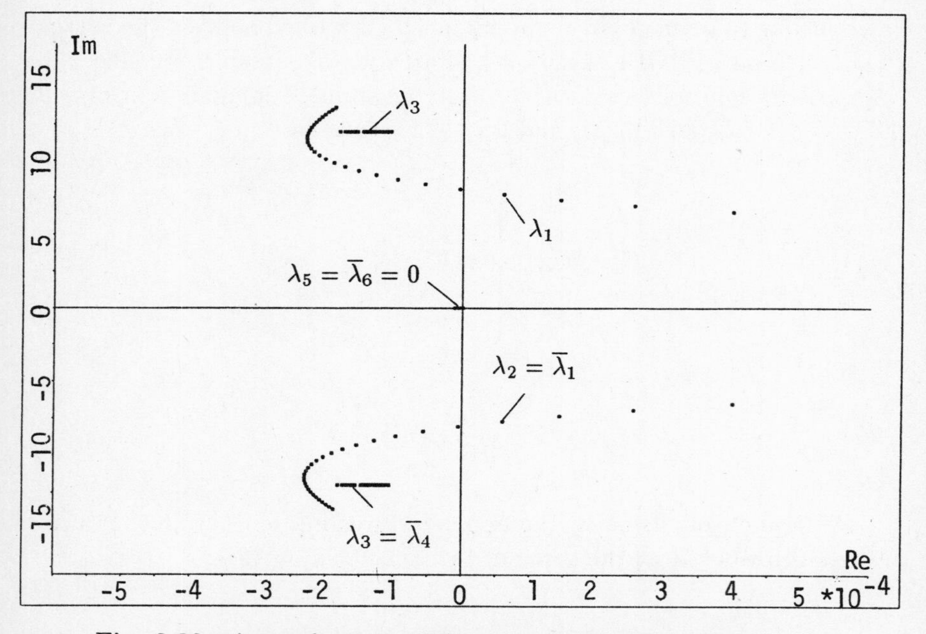

Fig. 2.38: eigenvalues of $D\mathbf{F}(\mathbf{Y}_0)$ in the complex plane $\mathbb{C}$

Besides two zero eigenvalues λ_5, λ_6 (the β-direction is free!) there is another set of pure negative real eigenvalues λ_7 and λ_8 of size 10^7 produced by the shear module G used in the model of the contact forces. These eigenvalues are not shown in fig. 2.38. Because of the large differences in the real parts of the eigenvalues the system becomes stiff. Hence, it is best to separate the ODE into its stiff part and into its nonstiff part if possible.

Immediately after the bifurcation the time of period T of the periodic solution is:

$$T = \frac{2\pi}{|\lambda_1(v_0^*)|} \approx 0.78\ s\ .$$

According to equation (2.108) T can be used as the initial end time for the numerical integration.

Fig. 2.39 shows that the behavior of bifurcation does not change if the shear module G is varied. This property can be used to eliminate the stiffness of the system and apply a conventional boundary value problem solver. If the periodic solution for a small value of G is found, it can be used to find the periodic solution for the actual G by homotopy techniques.

According to the dataset (cf. appendix) we obtain $v_0^* \approx 19\frac{m}{s}$. The stationary state $\mathbf{Y}_0$ (for instance straight-forward run with a constant velocity) is stable for $v_0 < v_0^*$. This is also shown by the phase curve (fig. 2.40) of $(q_1, \dot{q}_1) := (\gamma, \dot{\gamma})$ computed for $v_0 = 18\frac{m}{s} < v_0^*$.

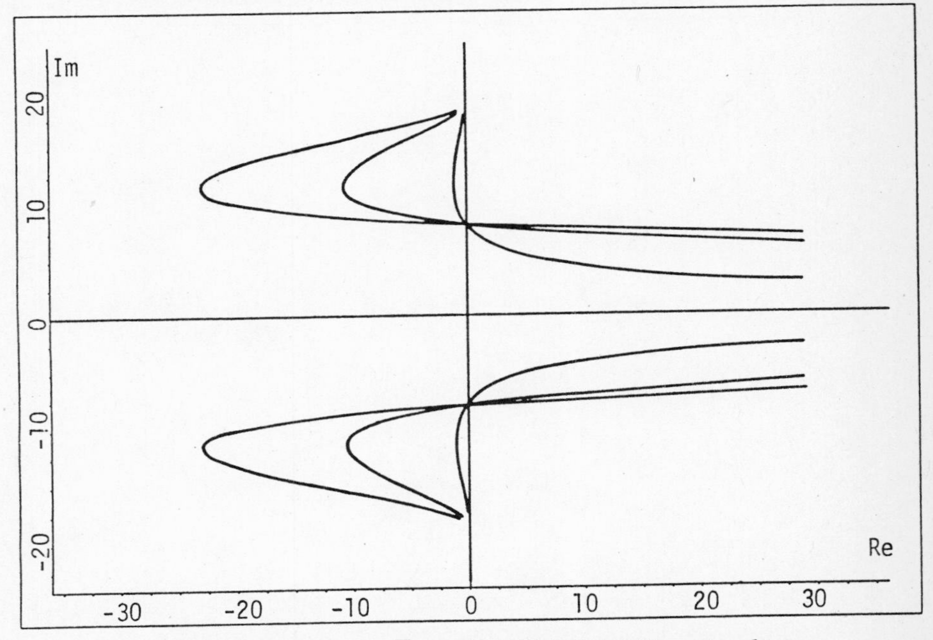

Fig. 2.39: eigenvalue $\lambda_1, \overline{\lambda}_1$ of $D\mathbf{F}(\mathbf{Y}_0)$ for different values of the shear module G

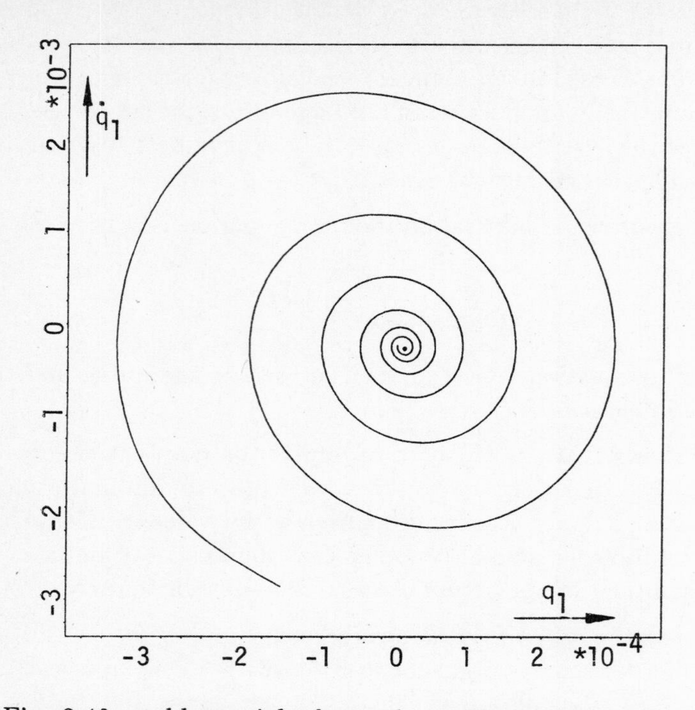

Fig. 2.40: stable straight-forward run $\mathbf{Y}_0$ represented for the coordinate $q_1 := \gamma$ according to fig. 2.1

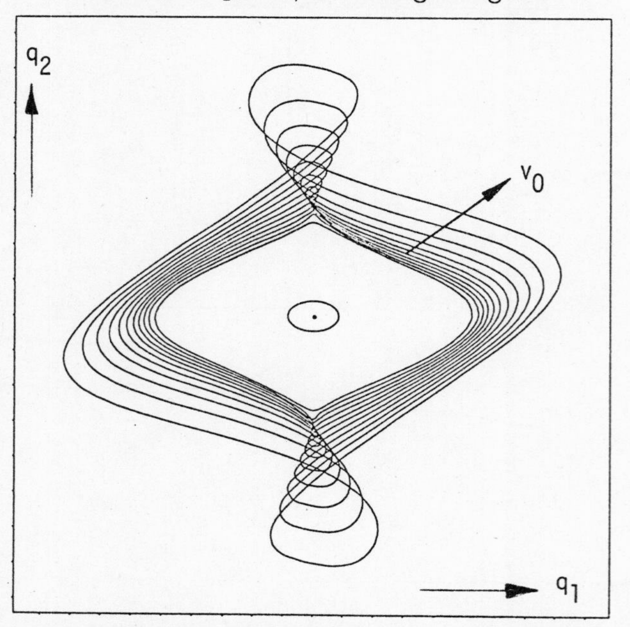

Fig. 2.41: behavior of the angles $(q_1, q_2) = (\gamma, \alpha)$ for $20\frac{m}{s} \leq v_0 \leq 30\frac{m}{s}$

For values $v_0 > v_0^*$ the expected limit cycle motion arises. The behavior for angles γ and α is shown in fig. 2.41. Each cycle belongs to a certain velocity $v_0 = v_0^* + k\Delta v, \quad k = 0, \ldots, 11, \quad \Delta v = 1\frac{m}{s}$.

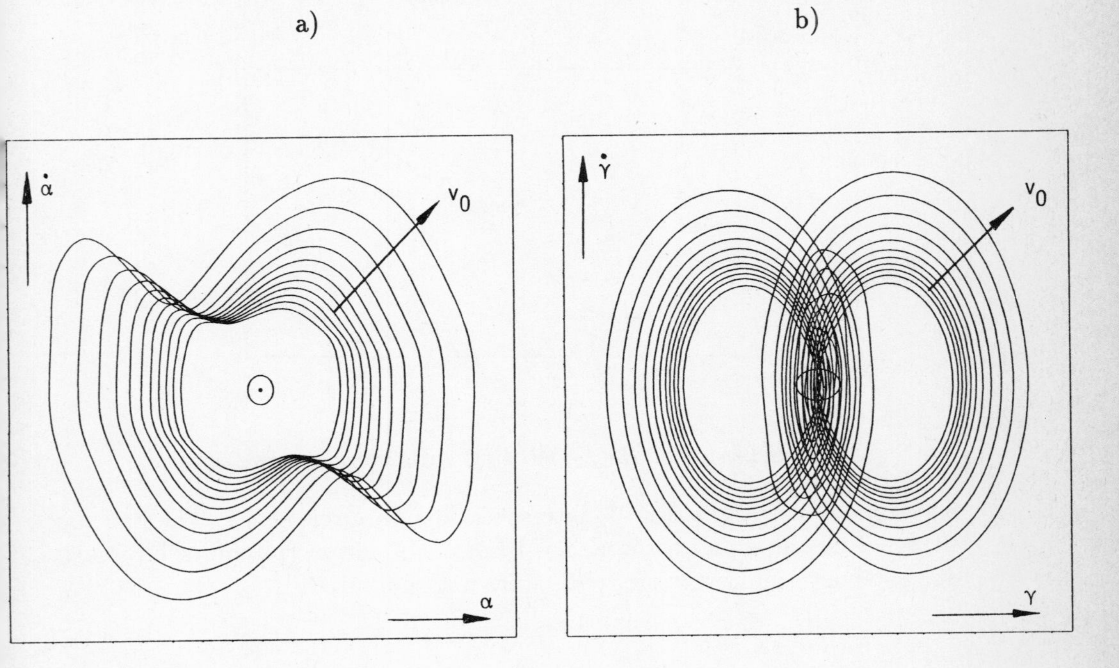

Fig. 2.42 a,b: projection of the limit cycles into the $(q_1, \dot{q}_1) = (\gamma, \dot{\gamma})$-plane and into the $(q_2, \dot{q}_2) = (\alpha, \dot{\alpha})$-plane

Fig. 2.42 a shows the phase curve for lateral deviation γ or $x = \frac{1}{\Gamma}\gamma$ respectively. Fig. 2.42 b represents the phase curve for yaw-angle α.

(b) Secondary HOPF-bifurcation

After the periodic solutions are computed, the monodromy matrix and its eigenvalues can be determined. Fig. 2.43 shows the behavior of the real part $Re(\lambda_2)$ of the second eigenvalue of the monodromy matrix (cf. also fig. 2.35) for limit cycles with $20\frac{m}{s} \leq v_0 \leq 40\frac{m}{s}$. While the conjugate complex pair of eigenvalues (λ_3, λ_4) stays on the unit circle with increasing energy or v_0 respectively, the eigenvalue λ_2 moves along the real axis and intersects the unit circle.

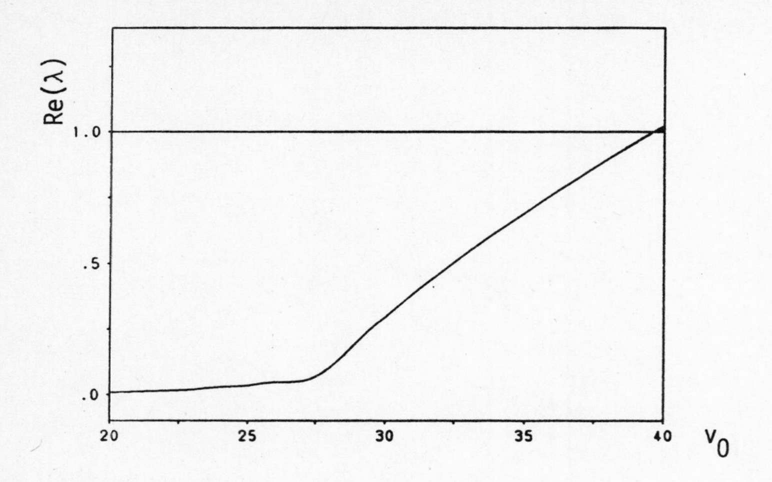

Fig. 2.43: behavior of $Re(\lambda_2(\mathbf{W}(T)))$ with increasing velocity v_0

At $v_0 \approx 40\frac{m}{s}$ the eigenvalue λ_2 intersects the unit circle at $+1$. The limit cycle becomes unstable and a secondary HOPF-bifurcation (pitchfork bifurcation) occurs. Further bifurcation points were not computed.

△

2.8.3 Necessary conditions for bifurcation

Periodic solutions of system (2.27) are (cf. chapter 2.6.1) solutions of the boundary value problem

$$\begin{aligned} \dot{\mathbf{x}} &= \mathbf{f}(\mathbf{x}, \mathbf{p}) \;, \\ \mathbf{0} &= \mathbf{r}(\mathbf{x}(T), \mathbf{x}(0)) := \mathbf{x}(T) - \mathbf{x}(0) \;. \end{aligned} \tag{2.157}$$

The period T is a function of the parameter $\mathbf{p} \in P$ ($\mathbf{p} \mapsto T(\mathbf{p})$). Hence, instead of $\mathbf{p}$ (or s if (cf. chapter 2.8.2) a curve $s \mapsto \gamma(s)$ inside P is given) we use the period T as bifurcation parameter. Each solution of the system (provided that the solution is LIPSCHITZ-continuous) is uniquely determined by its initial value $\boldsymbol{\xi} \in TM$. Therefore, equation (2.157) is equivalent to the computation of a point $\boldsymbol{\xi}$ intersected by a solution Φ_ξ such that

$$\mathbf{r}\,(\boldsymbol{\xi},\, \Phi_\xi(T)) = \mathbf{0} \;. \tag{2.158}$$

If $\mathbf{x}_0$ is a fix point of the dynamical system, $\boldsymbol{\xi} := \mathbf{x}_0$ and $\Phi_{\mathbf{x}_0}(t) = \mathbf{x}_0 \;\forall\; t$ is obviously a solution of equation (2.158).

According to the implicit function theorem, equation (2.158) has a unique solution $\boldsymbol{\xi}$ if the rank of the JACOBIan matrix $D_\xi \mathbf{r}(\boldsymbol{\xi}, \Phi_\xi(T))$ is maximal, that is if

$$det\,(D_\xi \mathbf{r}(\boldsymbol{\xi}, \Phi_\xi(T))) \neq 0 \ . \tag{2.159}$$

Hence, equation (2.159) is a necessary condition for the bifurcation of a fixed point $\mathbf{x}_0$ to a nontrivial periodic solution Φ_ξ.

If Φ_ξ is a nontrivial periodic solution, that is, if

$$\dot{\Phi}_\xi(t) = \mathbf{f}(\Phi_\xi(t), \mathbf{p}) \ , \tag{2.160}$$

then

$$\begin{array}{rcl} \Rightarrow \ D_\xi\left(\dot{\Phi}_\xi(t)\right) & = & D\mathbf{f}(\Phi_\xi(t), \mathbf{p}) \cdot D_\xi \Phi_\xi(t) \ . \\ \| & & \\ (D_\xi \Phi_\xi(t))^\bullet & & \end{array} \tag{2.161}$$

$$\begin{aligned} \Rightarrow \ D_\xi \Phi_\xi(t) &= \mathbf{W}(t) \cdot D_\xi \Phi_\xi(0) \\ &= \mathbf{W}(t) \qquad (\Phi_\xi(0) = \boldsymbol{\xi}) \ . \end{aligned} \tag{2.162}$$

$\mathbf{W}(t)$ is the WRONSKIan matrix of the time variant linearized system around the periodic solution Φ_ξ. Finally we obtain

$$\begin{aligned} D_\xi \mathbf{r}(\boldsymbol{\xi}, \Phi_\xi(T)) &= D_\xi(\Phi_\xi(T) - \boldsymbol{\xi}) \ , \\ &= \mathbf{W}(T) - \mathbf{E} \end{aligned} \tag{2.163}$$

For the trivial periodic solution $\Phi_{\mathbf{x}_0} = \mathbf{x}_0$ the WRONSKIan $\mathbf{W}(T)$ is the fundamental system of the linearized system

$$\dot{\mathbf{y}} = D\mathbf{f}(\mathbf{x}_0, \mathbf{p})\,\mathbf{y} \tag{2.164}$$

around the fixed point $\mathbf{x}_0$, that is

$$\begin{aligned} \mathbf{W}(T) &= e^{D\mathbf{f}(\mathbf{x}_0, \mathbf{p})T} \ , \\ &= \mathbf{X}\, e^{\Lambda T} \mathbf{X}^{-1} \ , \end{aligned} \tag{2.165}$$

where

$$\begin{aligned}
\boldsymbol{\Lambda} &:= diag\ \{\lambda_1, \ldots, \lambda_{2f}\}\ , \\
\mathbf{X} &:= [\mathbf{x}_1, \ldots, \mathbf{x}_{2f}]\ , \\
\lambda_i &: \quad i-\text{th eigenvalue of } D\mathbf{f}(\mathbf{x}_0, \mathbf{p})\ , \\
\mathbf{x}_i &: \quad \text{eigenvector corresponding to eigenvalue } \lambda_i\ .
\end{aligned}$$

$$\Rightarrow \quad det\left(D_{\xi}\mathbf{r}\left(\boldsymbol{\xi}, \Phi_{\xi}(T)\right)\right) = \prod_{i=1}^{2f}\left(e^{\lambda_i T} - 1\right)\ . \tag{2.166}$$

Hence, a bifurcation of a fixed point to a periodic solution can only occur if at least one eigenvalue λ_k becomes zero or pure imaginary, that is if $\lambda_k = i\omega_k$ with $\omega_k \in \mathbb{R}$. Immediately after the birfurcation the periodic solution has the period

$$T = \frac{2\pi}{\omega_k}\ . \tag{2.167}$$

In case of $\lambda_k = 0$, $\mathbf{x}_0$ is a degenerated fixed point which can be investigated by methods represented in chapter 2.4.

Equation (2.165) and (2.167) are an interesting supplement to the theorem of HOPF. Furthermore it is another proof of stability properties mentioned in the last chapter, since the results can be transferred to the bifurcation of periodic solutions. In order to do that, the following steps are necessary: $\mathbf{W}(T)$ has at least one 1-eigenvalue and $\mathbf{f}(\boldsymbol{\xi}, \mathbf{p})$ is eigenvector corresponding to the 1-eigenvalue. Hence the determinant in equation (2.159) is always zero. This is not surprising since in equation (2.158) the invariance of translation with respect to the initial value $\boldsymbol{\xi}$ is not taken into account. That means equation (2.158) has not a unique solution $\boldsymbol{\xi}$. One way to eliminate that problem is to define a POINCARÉ section Ω transversally to the periodic solution Φ_ξ. The intersecting point $\boldsymbol{\xi}$ of Φ_ξ with Ω is a fixed point of the POINCARÉ mapping $P : \Omega \rightarrow P(\Omega)$. According to the implicit function theorem P has a unique solution $\boldsymbol{\xi}$ if the JACOBIan of

$$\mathbf{r}(\boldsymbol{\xi}, P(\boldsymbol{\xi})) := P(\boldsymbol{\xi}) - \boldsymbol{\xi} \tag{2.168}$$

has full rank, that is (because of $dim\Omega = 2f - 1$) if

$$rk\left(D_{\xi}P(\boldsymbol{\xi}) - \mathbf{E}\right) = 2f - 1\ . \tag{2.169}$$

The definition of P says that $P(\boldsymbol{\xi}) = \Phi_\xi(T)$. Hence, equation (2.162) leads to

$$D_{\xi}P(\boldsymbol{\xi}) = \mathbf{W}(T) \tag{2.170}$$

or, if equation (2.159) is employed

$$rg(\mathbf{W}(T) - \mathbf{E}) = 2f - 1 \ . \tag{2.171}$$

That means bifurcation of a periodic solution Φ_ξ can only occur if a second 1-eigenvalue of $\mathbf{W}(T)$ exists.

$\boldsymbol{\xi}$ is also a fixed point of the mapping $P^n : \Omega \to P^n(\Omega), \quad n \in \mathbb{Z} - \{0\}$. Therefore equation (2.159) must hold for P^n too.

$$D_\xi P^n(\boldsymbol{\xi}) = (\mathbf{W}(T))^n \tag{2.172}$$

leads to the necessary condition for bifurcation:

$$rg\,(\mathbf{W}(T)^n - \mathbf{E}) = 2f - 1 \tag{2.173}$$

for all $n \in \mathbb{N} - \{0\}$. In other words: for an eigenvalue λ of $\mathbf{W}(T)$ defined by

$$\lambda = e^{i2\pi\left(\frac{m}{n}\right)} \qquad m, n \in \mathbb{Z} - \{0\} \tag{2.174}$$

a bifurcation may take place.

Each (irrational) point on the unit circle of $\mathbb{C}$ can be expressed as limit value of a sequence

$$\left(e^{i2\pi\left(\frac{m_k}{n_k}\right)}\right)_{k \in \mathbb{N}} \qquad \text{with} \quad m_k, n_k \ \in \ \mathbb{Z} - \{0\} \ . \tag{2.175}$$

That means intersection of an eigenvalue with the unit circle of $\mathbb{C}$ is a necessary condition for bifurcation of a periodic solution.

This result leads to another interesting statement: usually the POINCARÉ section Ω – in general – is a differentiable manifold and therefore locally isomorphic to $\mathbb{R}^{n-1}$. According to equation (2.137) a periodic solution is equivalent to a fixed point $\mathbf{x}_0 \in \Omega$ of the mapping P. Hence – according to chapter 2.2.2 – there is a local transformation of the coordinates $\mathbf{z} = Q(\mathbf{x})$, that transforms the non-linear POINCARÉ mapping $\mathbf{x} \mapsto P(\mathbf{x})$ into its linear part $\mathbf{z} \mapsto DP(\mathbf{x}_0)\,\mathbf{z}$ provided that the eigenvalues λ_k of $DP(\mathbf{x}_0)$ are not resonant (cf. equation (2.65)).

That is λ_k is resonant if a $\mathbf{m} \in \mathbb{N}^{n-1}$ with $|\mathbf{m}| \geq 2$ exists so that

$$\lambda_k = \boldsymbol{\lambda}^{\mathbf{m}} \ . \tag{2.176}$$

Or, because of $\lambda_k = r_k\, e^{i\omega_k}$, if

$$r_k = \mathbf{r}^{\mathbf{m}} := r_1^{m_1} \cdot \ldots \cdot r_{n-1}^{m_{n-1}} \tag{2.177}$$

and

$$\omega_k = \langle \mathbf{m}\,,\omega \rangle := \sum_{i=1}^{n-1} m_i\omega_i \quad ^{11} \ . \tag{2.178}$$

Hence, each eigenvalue located on the unit circle of $\mathbb{C}$ is obviously resonant since $DP(\mathbf{x}_0)$ is a real matrix. The definition $\lambda_1 := \lambda$ and $\lambda_2 := \overline{\lambda}$ leads to

$$\omega_1 = m_1\omega_1 + m_2\omega_2 \ . \tag{2.179}$$

Equation (2.179) is true for $\omega_2 = -\omega_1$ and any number $m_1 \geq 2$ and $m_2 = m_1 - 1$.

Condition (2.178) is automatically fulfilled if λ is located on the unit circle ($\mid \lambda \mid = 1!$).

In case all eigenvalues lie inside the unit circle, that is the periodic solution is stable, and in case the eigenvalues are not resonant the non-linear system will be equivalent to the linearized system around the periodic solution. λ becomes resonant if at least one eigenvalue λ intersects transversally the unit circle, for instance, by variation of parameters. The periodic solution loses its stability and a bifurcation may take place. At the same time, the transformation into normal form cannot be maintained. Law and order disappears or is transferred into a more complex form, what is more or less the same.

2.8.4 Stability and bifurcation in HAMILTONian systems

In foregoing investigations about stability we constantly assumed the dynamical system to be excited and/or damped. This assumption is not true for conservative or – in gerneral – HAMILTONian systems. Nevertheless fixed points and/or periodic solutions may also exist in HAMILTONian systems as the example of the double pendulum (chapter 2.5.4) shows.

If $\mathbf{z}_0$ is a fixed point of the HAMILTONian system (2.89) the linearized system around $\mathbf{z}_0$ is described by

$$\dot{\mathbf{y}} = \mathbf{J}\, D^2H\,(\mathbf{z}_0, \mathbf{p})\ \mathbf{y} \ . \tag{2.180}$$

Because of

$$det\left(\mathbf{J}\, D^2H\,(\mathbf{z}_0, \mathbf{p}) - \lambda E\right) = 0 \Longleftrightarrow det\left(\mathbf{J}\, D^2H\,(\mathbf{z}_0, \mathbf{p}) + \lambda E\right) = 0 \quad ^{12} \tag{2.181}$$

$-\lambda$ is an eigenvalue if λ is one.

[11]By the way, this condition contains all combinations of resonances as known from the classical theory of vibrations.

[12]For each real symmetric matrix $\mathbf{A}$ and symplectic matrix $\mathbf{J}$ ($\mathbf{J}^2 = -\mathbf{E}$ and $\mathbf{J}^{-1} = \mathbf{J}^T$) applies
$det(\mathbf{JA} - \lambda\mathbf{E}) = 0 \ \Leftrightarrow det\left((\mathbf{JA})^T - \lambda\mathbf{E}\right) = 0 \quad \Leftrightarrow \quad det\left(\mathbf{A}^T\mathbf{J}^T - \lambda\mathbf{E}\right) = 0$,
$\Leftrightarrow det\left(\mathbf{AJ}^T - \lambda\mathbf{E}\right) = 0 \quad \Leftrightarrow \quad det\left[\mathbf{J}\left(\mathbf{AJ}^T - \lambda\mathbf{E}\right)\mathbf{J}\right] = 0$,
$\Leftrightarrow det\left(\mathbf{JA} - \lambda\mathbf{J}^2\right) = 0 \quad \Leftrightarrow \quad det\left(\mathbf{JA} + \lambda\mathbf{E}\right) = 0$.

A consequence of this property is that a fix point $\mathbf{z}_0$ is unstable if the linearized system around $\mathbf{z}_0$ has eigenvalues with not vanishing real parts. Hence, a HAMILTONian system has only "limit stable" or unstable fix points. Limit stability means that all eigenvalues are conjugate imaginary. According to equation (2.167) limit stable fix points bifurcate to periodic solutions and the period T (immediately after bifurcation) results from the eigenfrequencies of the linearized system around $\mathbf{z}_0$.

In other words, a bifurcation of a limit stable fix point $\mathbf{z}_0$ of a HAMILTONian system leads always to a periodic eigenmode of the linearized system around $\mathbf{z}_0$. This kind of bifurcation can be considered as primary HOPF-bifurcation of a HAMILTONian system. The evolution of periodic solutions after bifurcation of the fix point $\mathbf{z}_0 = \mathbf{0}$ is shown in case 1 of chapter 2.5.4.

The stability of a periodic solution Φ_ξ is determined by the distribution of the eigenvalues of the monodromy matrix according to equation (2.153). For HAMILTONian systems $\frac{1}{\lambda}$ is an eigenvalue of $\mathbf{W}(T)$ if λ is an eigenvalue (cf. [ARNOLD 1988a]). Hence, either the eigenvalues lie on the unit circle K_1 of $\mathbb{C}$ ($|\lambda| = 1$) or they are symmetric to the unit circle K_1.

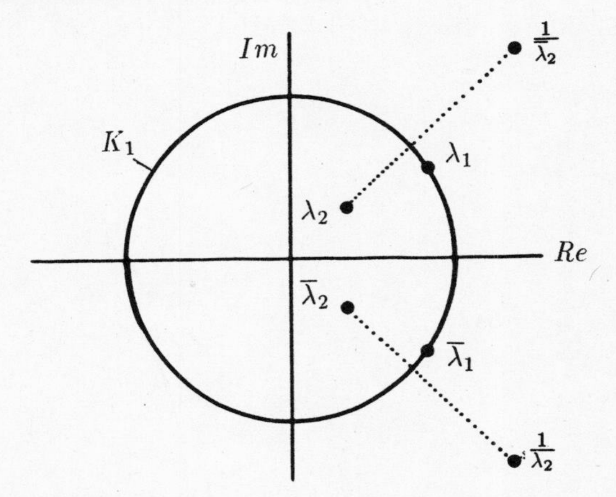

Fig. 2.44: distribution of the eigenvalues of the monodromy matrix of a periodic solution of a HAMILTONian system

An eigenvalue λ on K_1 can only leave K_1 if λ and $\overline{\lambda}$ meet on the real axis or λ meets another eigenvalue on K_1 (cf. fig. 2.45).

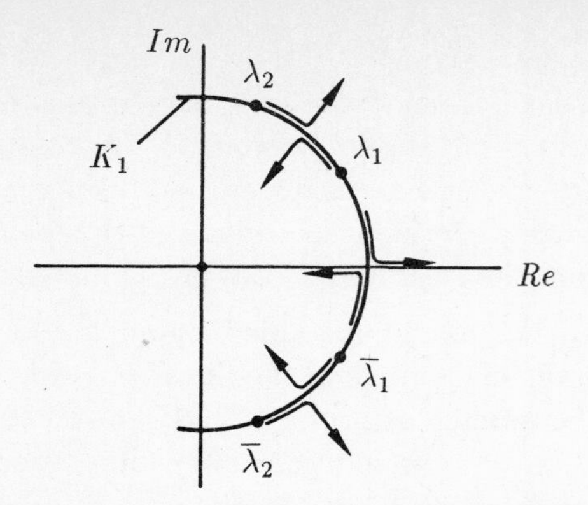

Fig. 2.45: options for an eigenvalue λ to leave the unit circle K_1 of $\mathbb{C}$

In HAMILTONian systems the monodromy matrix $\mathbf{W}(T)$ has always two 1-eigenvalues. Let Π be the two dimensional manifold of periodic solutions produced by variation of the energy parameter h, an arbitrary curve $\Gamma : [h_1, h_2] \to \Pi$ can be defined on Π that passes the periodic solutions transversally. The set of intersection points of Γ and the periodic solutions define initial points of the periodic solutions (cf. fig. 2.46).

That is

$$\Gamma(h) := \Phi_{\Gamma(h)}(0) \ . \tag{2.182}$$

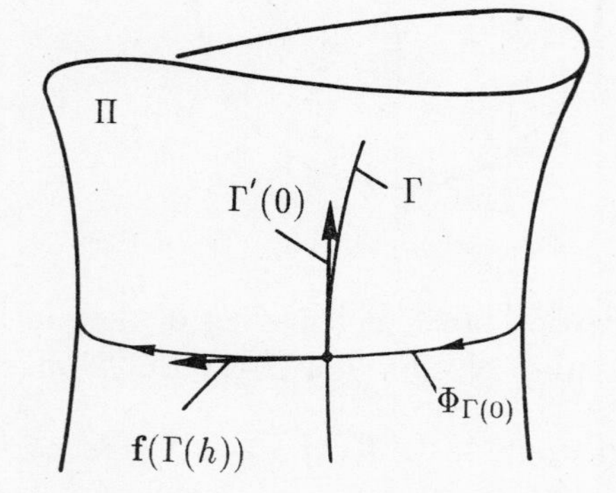

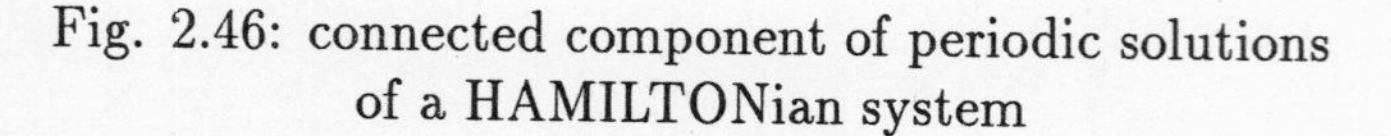

Fig. 2.46: connected component of periodic solutions of a HAMILTONian system

Hence, for each $h \in [h_1, h_2]$, $\Phi_{\Gamma(h)}$ is a solution, that is

$$\dot{\Phi}_{\Gamma(h)}(t) = \mathbf{f}\left(\Phi_{\Gamma(h)}(t), \mathbf{p}\right) \ . \tag{2.183}$$

$$\Rightarrow \frac{d}{dh}\dot{\Phi}_{\Gamma(h)}(t) = D\mathbf{f}\left(\Phi_{\Gamma(h)}(t), \mathbf{p}\right) \cdot \frac{d}{dh}\Phi_{\Gamma(h)}(t) \; . \tag{2.184}$$

$$\Vert$$

$$\left[\frac{d}{dh}\Phi_{\Gamma(h)}(t)\right]^{\bullet}$$

$$\Rightarrow \frac{d}{dh}\Phi_{\Gamma(h)}(t) = \mathbf{W}(t) \cdot \frac{d}{dh}\Phi_{\Gamma(h)}(0) \; . \tag{2.185}$$

Remember that $\Phi_{\Gamma(h)}(0) = \Phi_{\Gamma(h)}(T) = \Gamma(h)$.

$$\Rightarrow \Gamma'(h) = \mathbf{W}(T)\Gamma'(h) \; ; \quad \left(' := \frac{d}{dh}\right) \; . \tag{2.186}$$

That means $\Gamma'(h)$ is an eigenvector of $\mathbf{W}(T)$ corresponding to the eigenvalue 1 and – because of the transversality of Γ and Φ_Γ – the vectors $\Gamma'(h)$ and $\mathbf{f}(\Gamma(h))$ are linear independent.

According to equation (2.174) the presence of the second 1-eigenvalue means that a POINCARÉ mapping $P : \Omega \rightarrow \Omega$ (Ω, transversal to the periodic solutions) fulfills always the necessary condition for bifurcation. Nevertheless the periodic solution keeps its stability since the eigenvalues can only leave the unit circle on the real axis. In that case stability means that a small disturbance either diverges or converges to the periodic solution. In other words: only definition (a) of chapter 2.8 is true. Therefore, periodic solutions of HAMILTONian systems are either limit stable or unstable.

∇

Example 2.23:

We want to investigate the periodic solutions of the double pendulum (chapter 2.5.4) with respect to stability. We apply the algorithm described in equation (2.155).

The figures 2.47 and 2.48 show the result for case 1 and for case 2 (chapter 2.5.4). The pictures illustrate the behavior of the real part and imaginary part of the non 1-eigenvalue of the monodromy matrix with increasing value of the HAMILTONian function H.

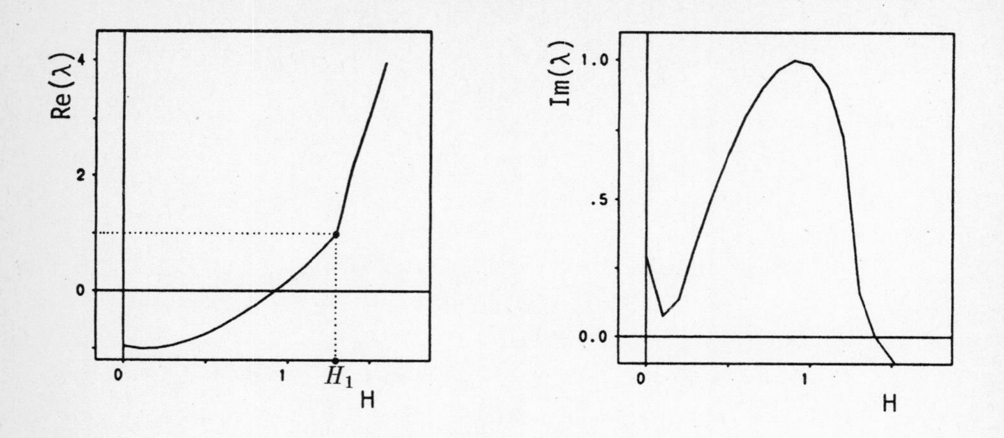

Fig. 2.47: behavior of the non 1-eigenvalue of $\mathbf{W}(T)$
for case 1, chapter 2.5.4

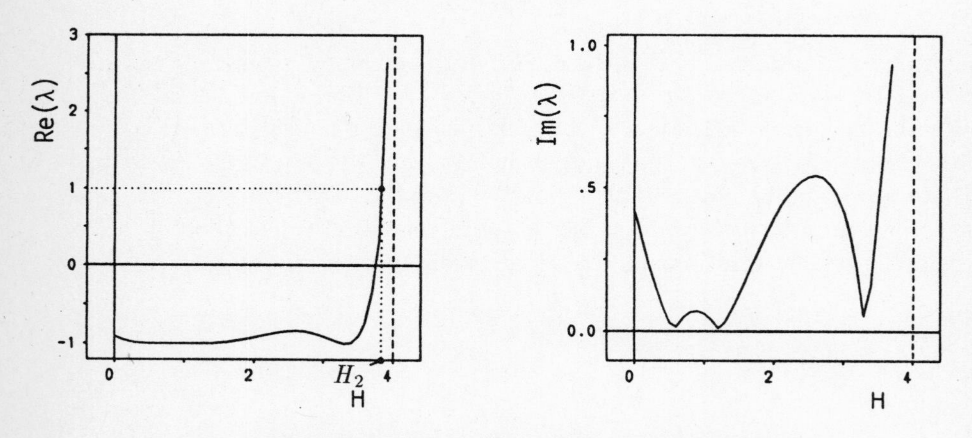

Fig. 2.48: behavior of the non 1-eigenvalue of $\mathbf{W}(T)$
for case 2, chapter 2.5.4

The periodic solution of case 1 becomes already unstable for $h = h_1 = \ldots < 2$, while the stability of the periodic solution of case 2 is stable up to value $h_2 \approx 4.0$.

Fig. 2.49 shows the stable and unstable branches of all periodic (k_1, k_2)-solutions that exist in the energy range $[0,\ h_4]$. The numbers correspond to the dataset given in the appendix.

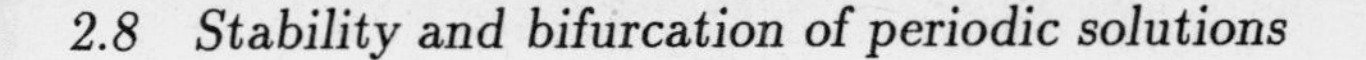

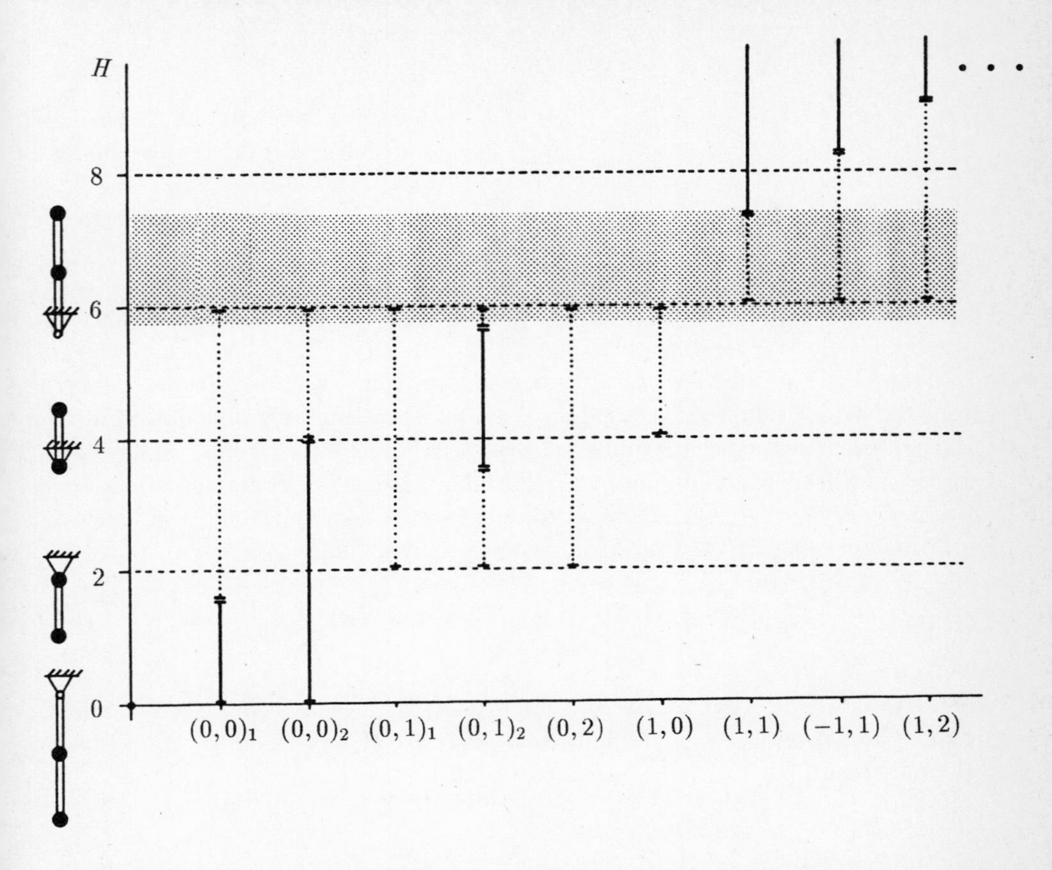

Fig. 2.49: stable (–) and unstable (· · ·) branches of the periodic (k_1, k_2)-solutions

The shaded zone shows the energy range where no stable periodic solution exists. The existence of such a range is the reason that the behavior of the double pendulum is very complex or chaotic respectively for energy values inside that range.

If the energy is increased, the periodic solutions become stable again. This is plausible since for high energy values both pendulums are rotating and the centrifugal forces lead to stable rotations.

△

3 Differentiable dynamical systems with discontinuities

Auch das Zufälligste ist nur ein auf entfernterem Wege herangekommenes Notwendiges.

A. Schopenhauer

Mainly in technical applications it is often necessary or convenient to use dynamical models where the corresponding vector field $\mathbf{f}$ (cf. equation (2.1)) has discontinuities with respect to the state space variables $\mathbf{x} \in TM$. Even if the discontinuities occur very frequently it is useful to investigate a dynamical system based on a deterministic description. Statistical methods and stochastic techniques – particularly for systems with discontinuities – can be found in [KARAGIANNIS 1989]. Since the vector field $\mathbf{f}$ is almost everywhere – that is except for a zeroset – differentiable, it is useful to separate the state space TM into disjunct open subsets X_i:

$$TM = \overline{\bigcup_{i \in I} X_i} \quad , \qquad (I := \text{ indexset}) \tag{3.1}$$

For each subset X_i the dynamical system is described by a differentiable field $\mathbf{f}_i$ according to equation (2.1). If the indexset I is finite (for instance $| I |= N$), then we want to define $\mathbf{f}$ by

$$\mathbf{f} = \begin{cases} \mathbf{f}_1 & \text{if } \mathbf{x} \in X_1 \ , \\ \vdots & \\ \mathbf{f}_N & \text{if } \mathbf{x} \in X_N \end{cases} \tag{3.2}$$

The system (3.2) is called "structural variant" if the functions $\mathbf{f}_i$ are different.

∇

Example 3.1: (structural invariant system with impact and backlash)

The rattling model of a one staged gear, investigated by [PFEIFFER 1988a], is equivalent to the model shown in fig. 3.1.

A small ball (mass m) is moving in a pan (mass $M \gg m$, length v). The mass is damped (viscous damping, coefficient δ). The pan moves according to the excitation function $e(t, p_i)$ with respect to an inertial frame.

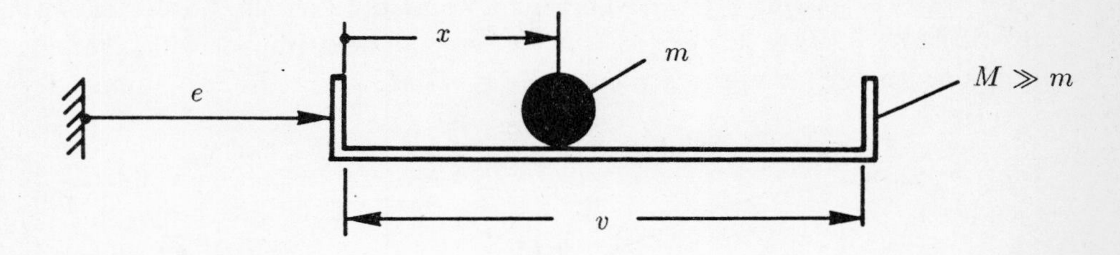

Fig. 3.1: rattling model with 1 degree of freedom

x is the relative coordinate between pan and ball. The equation of motion for the ball is

$$\ddot{x} + \left(\frac{\delta}{m}\right) \dot{x} = -\ddot{e}(t, p_i) \ ,$$

or, transformed into a autonomous first order ODE ($x_1 := x, \ \ x_2 := \dot{x}, \ \ x_3 := t$):

$$\underbrace{\begin{array}{l} \dot{x}_1 \\ \dot{x}_2 \\ \dot{x}_3 \end{array}}_{\dot{\mathbf{x}}} \begin{array}{c} = \\ = \\ = \end{array} \underbrace{\begin{array}{l} x_2 \ , \\ -\ddot{e}(x_3, p_i) - \left(\frac{\delta}{m}\right) x_2 \ , \\ 1 \ . \end{array}}_{\mathbf{f(x,p)}}$$

$\mathbf{p} := \left(\left(\frac{\delta}{m}\right) \ , \ p_2 \ , \ p_3 \ , \ \ldots\right)$ is the parameter vector.

The state space $TM = [0 \ , \ v] \times \mathbb{R}^2$ is bounded with respect to the relative coordinate x_1. If the ball hits one of the side walls of the pan, an impact will occur if the velocity x_2 is not zero. The impact model and the state $\mathbf{x}^-$ immediately before impact determine the state $\mathbf{x}^+$ immediately after impact.

If Φ_ξ is the trajectory of $\dot{\mathbf{x}} = \mathbf{f}(\mathbf{x}, \mathbf{p})$ with initial point $\boldsymbol{\xi} = (\xi_1, \xi_2, \xi_3) \in TM$ where $\Phi_\xi(\xi_3) = \boldsymbol{\xi}$, and $t_s > \xi_3$ is the time of impact,

$$\lim_{\substack{t \to t_s \\ t < t_s}} \Phi_\xi(t) \neq \lim_{\substack{t \to t_s \\ t > t_s}} \Phi_\xi(t) \ .$$

holds in case the impact model

$$\mathbf{x}^+ = diag\{1 \ , \ -\varepsilon \ , \ 1\} \, \mathbf{x}^- \ ; \quad 0 \leq \varepsilon \leq 1$$

is used.

We obtain a unique solution Φ_ξ if $\mathbf{f}$ is LIPSCHITZ–continuous and the impact model includes a unique connection between the state before and after impact for each $\boldsymbol{\xi} \in TM$. In that case, the trajectory has the same number of discontinuities as impacts occur.

Furthermore to each point $\boldsymbol{\xi} \in TM$ belongs a point $\mathbf{b}(\boldsymbol{\xi})$ on the border or better separatrix $\partial TM := \{0\} \times \mathbb{R}^2 \cup \{v\} \times \mathbb{R}^2$ where the next impact will take place. Corresponding to $\mathbf{b}(\boldsymbol{\xi})$ there is a point $\mathbf{a}(\boldsymbol{\xi}) \in \partial TM$ where the last impact took place (cf. fig. 3.2).

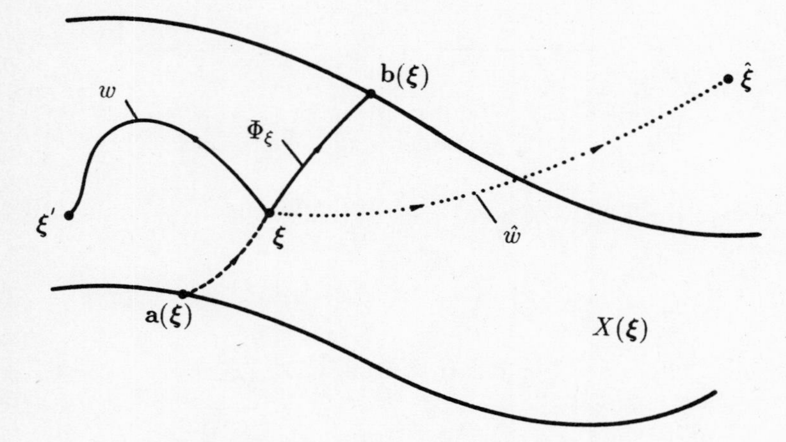

Fig. 3.2: definition of the subsets $X(\boldsymbol{\xi})$

The state space TM can be partitioned into open subsets as follows: if $\boldsymbol{\xi} \in TM$ is arbitrary,

$$X(\boldsymbol{\xi}) := \{\boldsymbol{\xi}' \in TM \big| \ \exists \, \text{curve } w : [0,1] \to TM - \partial TM \text{ with } w(0) = \boldsymbol{\xi}, w(1) = \boldsymbol{\xi}'\}$$

is the largest open subset of TM that contains $\boldsymbol{\xi}$ and where all trajectories are differentiable.

The separatrix of $\overline{X(\boldsymbol{\xi})}$, that is

$$\partial\overline{X(\boldsymbol{\xi})} = A \cup B \qquad \subset \partial TM$$

with

$$A := \{\mathbf{a}(\boldsymbol{\eta}) \mid \boldsymbol{\eta} \in X(\boldsymbol{\xi})\},$$

$$B := \{\mathbf{b}(\boldsymbol{\eta}) \mid \boldsymbol{\eta} \in X(\boldsymbol{\xi})\},$$

contains all points of the state space where impacts appear for trajectories in $X(\boldsymbol{\xi})$. On the border the vector field $\mathbf{f}$ is discontinuous. Hence,

$$TM = \overline{\bigcup_{\xi \in TM} X(\boldsymbol{\xi})}$$

is, – according to equation (3.2) – the partition of the state space TM into disjunct open subsets $X(\boldsymbol{\xi})$.

<u>Remarks:</u>

(1) To put it more precisely, the impact model is an algebraic constraint which is – as above – only active on the separatrix ∂TM of the state space. Therefore, in addition to the dynamical system $\dot{\mathbf{x}} = \mathbf{f}(\mathbf{x}, \mathbf{p})$ the impact model is needed for a complete description.

(2) The point $\hat{\boldsymbol{\xi}} \in TM$, shown in fig. 3.2, does not lie on $X(\boldsymbol{\xi})$ since each curve from $\boldsymbol{\xi}$ to $\hat{\boldsymbol{\xi}}$ leads across a separatrix of discontinuities of TM.

(3) Each trajectory crosses all areas X of the partition of TM. The impacts are countable and therefore the partition is also countable.

(4) If the exciting function e is T-periodic, TM may also be considered as a "covering" with the frame

$$X := [0, v] \times \mathbb{R} \times [0, T] \ .$$

In that case TM is isomorph to the "trivial fibering" $X \times \mathbb{Z}$.

(5) The vector field $\mathbf{f}$ has the same structure on each subset X_i of the partition. Hence, according to the definition mentioned above, systems with impacts are structural invariant.

(6) Without any restriction, the description can be applied to rigid body systems with more degrees of freedom and backlash.

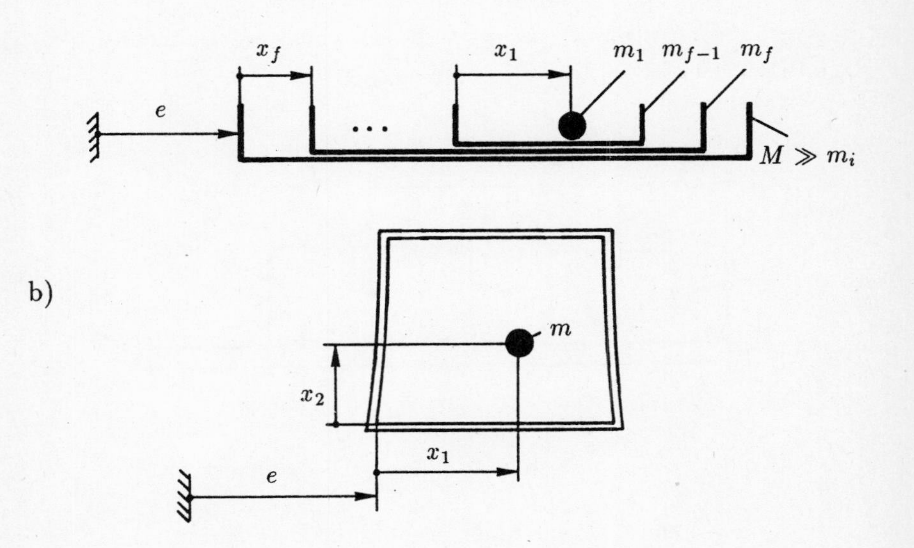

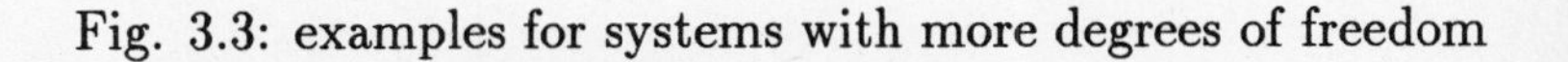
Fig. 3.3: examples for systems with more degrees of freedom

Fig. 3.3 a) shows a multiple-staged rattling model with f degrees of freedom and constant backlash ranges v_i. The state space can be expressed by

$$TM = \left(\mathop{\times}_{i=1}^{f} [0, v_i] \right) \times \mathbb{R}^f \times \mathbb{R} \ .$$

Fig. 3.3 b) represents the unloaded wheel of a one staged gear with axial play of the bearing. In that case the backlash v_1 of the cog depends on the position x_2. The axial play v_2 is constant. The configuration space is described by

$$M := \{(x_1, x_2) \mid a(x_2) \leq x_1 \leq b(x_2) \ ; \quad 0 \leq x_2 \leq v_2\} \ .$$

Hence, the state space is

$$TM = M \times \mathbb{R}^2 \times \mathbb{R} \ .$$

The corresponding impact model for systems with more degrees of freedom can be found, for instance, in [PFEIFFER 1984] or [PFEIFFER, KÜCÜKAY 1985]. Other applications such as impact models between rail and run of railway vehicles were investigated by [MEIJAARD, DE PATER 1989] (see also page 75).

△

▽

Example 3.2: (structural variant system with dry friction)

A block (mass m) lies on a plane (mass $M \gg m$) which moves according to the function $e(t, p_i)$ (fig. 3.4).

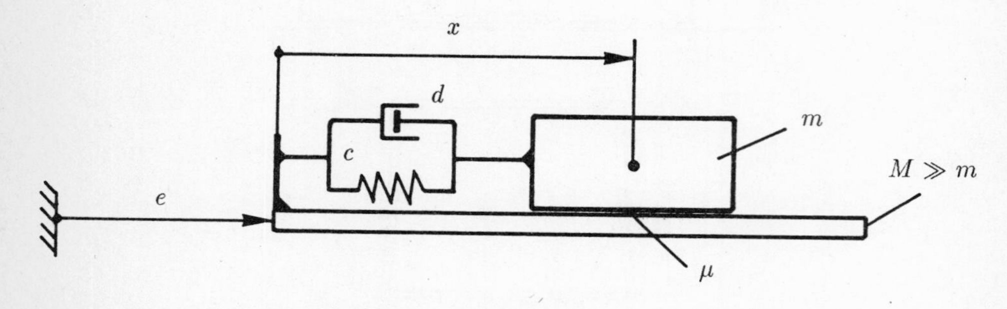

Fig. 3.4: block on a moving plane with dry friction

With respect to a plane reference, the block is mounted by a spring- and damping-element. The coordinate x_e marks the unloaded spring, x is the coordinate of the streched spring.

If the friction model is assumed as

$$\text{friction force} \leq \mu \cdot \text{ normal force} \quad ,$$

the following two cases have to be distinguished:

1. sticking

$$x(t) = const.$$

2. slipping

$$\ddot{x} + \left(\frac{\delta}{m}\right)\dot{x} + \left(\frac{c}{m}\right)x + \ddot{e}(t,p_i) + \mu g\, sgn(\dot{x}) = 0 \ .$$

Sticking takes place if

$$1) \qquad \dot{x} = 0 \ ,$$

$$2) \qquad \left|\left(\frac{c}{m}\right)x + \left(\frac{\delta}{m}\right)\dot{x} + \ddot{e}(t,p_i)\right| \leq \mu g$$

holds.

Hence, the state space $TM = \mathbb{R}^3$ with the coordinates $x_1 := x$, $x_2 := \dot{x}$ and $x_3 := t$ has to be separated into the subsets

$$X_1 := \left\{\mathbf{x} \in \mathbb{R}^3 \middle| x_2 < 0\right\}$$

$$X_2 := \left\{\mathbf{x} \in \mathbb{R}^3 \middle| x_2 > 0\right\}$$

$$G := \left\{\mathbf{x} \in \mathbb{R}^3 \middle| x_2 = 0 \ , \ \left|\left(\frac{c}{m}\right)x_1 + \ddot{e}(x_3,p_i)\right| > \mu g\right\}$$

$$H := \left\{\mathbf{x} \in \mathbb{R}^3 \middle| x_2 = 0 \ , \ \left|\left(\frac{c}{m}\right)x_1 + \ddot{e}(x_3,p_i)\right| \leq \mu g\right\} \ .$$

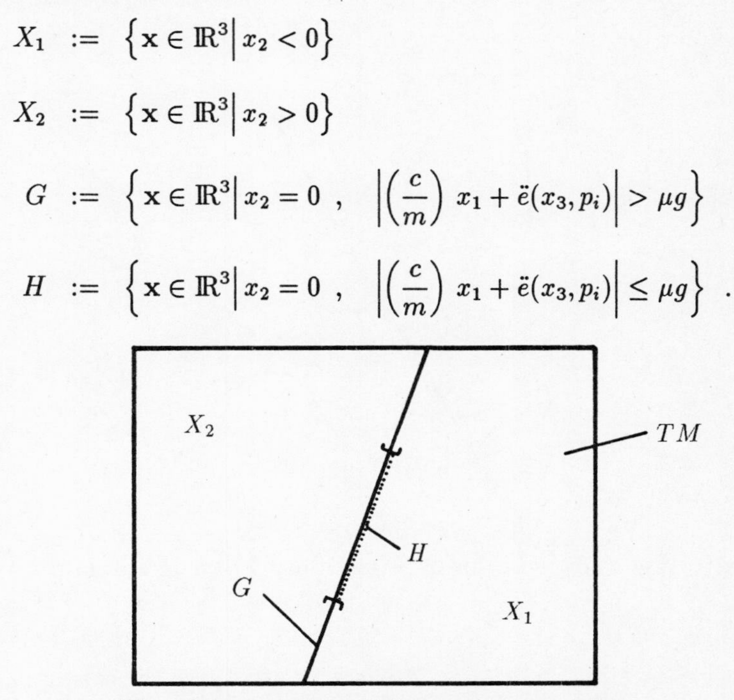

Fig. 3.5: partition of the state space TM

While H is representing the range of sticking, G just determines the set of discontinuities of the field $\mathbf{f}$.

According to the definition given above, the system is structural variant. Namely if $\mathbf{f}_+$ and $\mathbf{f}_-$ is defined by

$$\mathbf{f}_\pm(\mathbf{x},\mathbf{p}) := \begin{bmatrix} x_2 \\ -\left(\frac{c}{m}\right) x_1 - \left(\frac{\delta}{m}\right) x_2 - \ddot{e}(x_3, p_i) \mp \mu g \\ 1 \end{bmatrix} ,$$

the vector field $\mathbf{f}$ is given by

$$\mathbf{f}(\mathbf{x},\mathbf{p}) = \begin{cases} \mathbf{f}_+(\mathbf{x},\mathbf{p}) & \forall \quad \mathbf{x} \in X_1 \\ \mathbf{f}_-(\mathbf{x},\mathbf{p}) & \forall \quad \mathbf{x} \in X_2 \end{cases} .$$

Remarks:

(1) Any trajectory that comes in contact with H will stay on H as long as the separatrix between H and G is reached. At that point, the trajectory leaves G (cf. fig. 3.6).

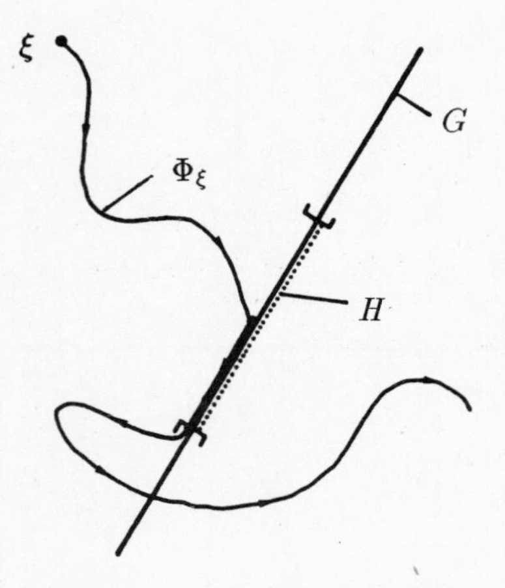

Fig. 3.6: behavior of a trajectory Φ_ξ on H and G

If Φ_ξ does not leave the range H of sticking, $\mathbf{x}(\boldsymbol{\xi}) := \lim_{t\to\infty} \Phi_\xi(t)$ is a fix point of the system, provided the limit exists. In that case the attractor is given by the subset

$$A := \left\{ \mathbf{x} \in H \,\middle|\, \left| \left(\frac{c}{m}\right) x_1 + \ddot{e}(x_3, p_i) \right| \le \mu g \quad \forall \ x_3 \in \mathbb{R} \right\} .$$

(2) There are two possibilities for Φ_ξ to hit G: Either Φ_ξ crosses G transversally or Φ_ξ touches G tangentially. A criterion for this is

$$\lim_{\substack{t \to t_A \\ t < t_A}} \mathbf{f}\,(\Phi_\xi(t_A), \mathbf{p}) \notin G \quad \Rightarrow \quad \Phi_\xi \text{ crosses } G \text{ transversally}$$

$$\lim_{\substack{t \to t_A \\ t < t_A}} \mathbf{f}\,(\Phi_\xi(t_A), \mathbf{p}) \in G \quad \Rightarrow \quad \Phi_\xi \text{ touches } G \text{ tangentially .}$$

t_A is the time at crossing or touching. The limes indicates the vector field ($\mathbf{f}_+$ or $\mathbf{f}_-$) that has to be used.

(3) Without any restriction, the description given above can be extended to more degrees of freedom and dry friction elements. If R_k is the k-th friction element (friction coefficient μ_k) between two bodies (i) and (j), $\mathbf{v}_{rel}(\mathbf{q}, \dot{\mathbf{q}}, t) \in T$ its relative velocity at the contact point K and $F_N(\mathbf{q}, \dot{\mathbf{q}}, t)$ the normal force acting at the contact point K, then

$$G_k := \{\mathbf{x} \in TM \mid \mathbf{v}_{rel}(\mathbf{x}) = \mathbf{0} \ , \ \| \mathbf{F}_Z(\mathbf{x}) \| > \mu_k \cdot F_N(\mathbf{x})\}$$

$$H_k := \{\mathbf{x} \in TM \mid \mathbf{v}_{rel}(\mathbf{x}) = \mathbf{0} \ , \ \| \mathbf{F}_Z(\mathbf{x}) \| \leq \mu_k \cdot F_N(\mathbf{x})\}$$

applies.

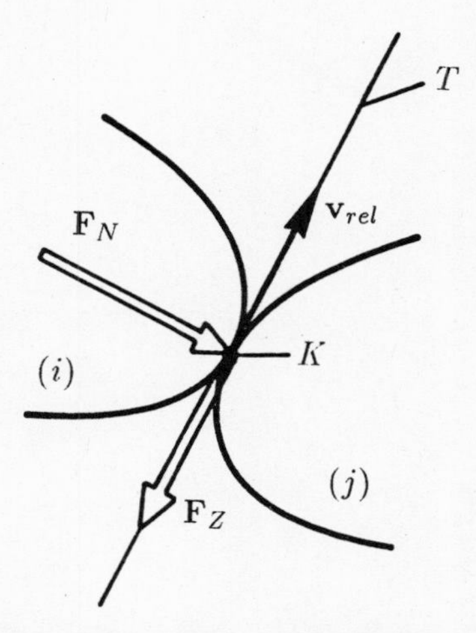

Fig. 3.7: friction element between two bodies (i) and (j)

The state is indicated by $\mathbf{x} := (\mathbf{q}, \dot{\mathbf{q}}, t)$. $\mathbf{F}_Z(\mathbf{x})$ is the reaction force acting in the contact surface T. $\mathbf{F}_Z(\mathbf{x})$ is a "passive" force for each state $\mathbf{x} \in H_k$ and an "active" force for any state $\mathbf{x} \in G_k$.

If the sets (zerosets) G_k and H_k are combined as follows:

$$G := \bigcup_{k=1}^{N} G_k \quad , \qquad H := \bigcup_{k=1}^{N} H_k \ ,$$

the partition of TM into the sets X_i – according to equation (3.1) – is determined by the unification $H \cup G$.

Notice that – in general – the sets $G_k \cup H_k$ are differentiable manifolds of the dimension $m := dim(TM) - 1$ and not – as above – linear spaces.

$\triangle$

3.1 The recursive description

The flow of dynamical systems with discontinuities crosses all sets X_i, $i \in I$ of the partition of TM. On each X_i the system can be studied according to chapter 2 without any restriction. The separatrix of any set X_i is a zeroset of TM or, more strictly spoken, is a manifold of dimension $n-1$.

This fact makes it possible to choose the separatrix or the manifold respectively as a "POINCARÉ section" of the dynamical system in a very convenient way. Namely, because of the presumed uniqueness of the trajectories in the phase space the knowledge of the "intersecting points" of the phase curves on the separatrices allows essential conclusions about the dynamical behavior of the system.

In order to determine the intersecting points we need information about the flow. That is, we have to compute solution of the system at least numerically.

The separatrix $\partial\overline{X}_i$ ($dim\, \partial\overline{X}_i = n-1$) of $\overline{X}_i$ is expressed by a zeroset of an algebraic function

$$g_i : TM \;\rightarrow\; \mathbb{R} \;. \tag{3.3}$$

For determination of the recursion, that is the POINCARÉ mapping

$$\boldsymbol{\Psi}_i \;:\; g_i^{-1}(0) \rightarrow g_i^{-1}(0) \;,$$

the following definition is useful:

Definition (3.1):

If $t^{(k)}$ is the time where a solution Φ_ξ intersects the separatrix $g_i^{-1}(0)$, then we define state immediately before the discontinuity by

$$\mathbf{x}_-^{(k)} := \lim_{\substack{t \rightarrow t^{(k)} \\ t < t^{(k)}}} \Phi_\xi(t) \tag{3.4}$$

and the state immediately after the discontinuity is defined by

$$\mathbf{x}_+^{(k)} := \lim_{\substack{t \rightarrow t^{(k)} \\ t > t^{(k)}}} \Phi_\xi(t) \tag{3.5}$$

(cf. fig. 3.8).

○

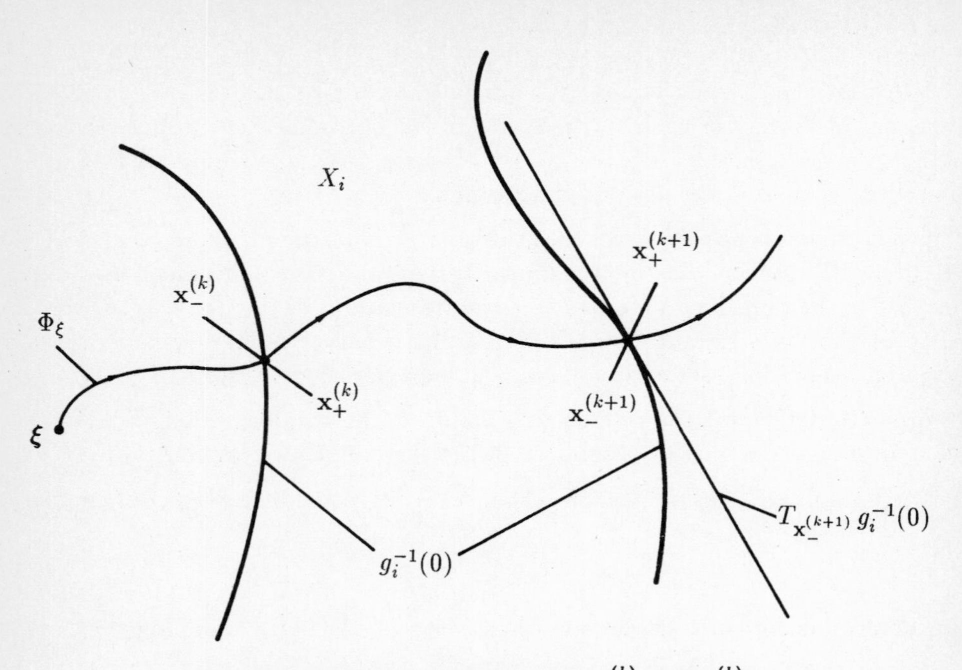

Fig. 3.8: definition of the points $\mathbf{x}_-^{(k)}$ and $\mathbf{x}_+^{(k)}$

The definition requires that $\mathbf{x}_-^{(k)} \in g_i^{-1}(0)$ as well as $\mathbf{x}_+^{(k)} \in g_i^{-1}(0)$.

The law, to connect $\mathbf{x}_-^{(k)}$ and $\mathbf{x}_+^{(k)}$ at the discontinuity is expressed by a differentiable function

$$\sigma_i : g_i^{-1}(0) \to g_i^{-1}(0) \ , \quad \mathbf{x}_-^{(k)} \mapsto \mathbf{x}_+^{(k)} \ . \tag{3.6}$$

The recursion $\boldsymbol{\Psi}_i$ maps state $\mathbf{x}_+^{(k)}$ into state $\mathbf{x}_+^{(k+1)}$:

$$\mathbf{x}_+^{(k+1)} = \boldsymbol{\Psi}_i \left(\mathbf{x}_+^{(k)} \right) \ . \tag{3.7}$$

If Φ_ξ intersects the separatrix $g_i^{-1}(0)$ transversally, then, provided $\mathbf{x}_+^{(k)}$ is known, $\boldsymbol{\Psi}_i$ can be computed by the boundary value problem according to equation (3.8).

For purpose of numerical computation of $\boldsymbol{\Psi}_i \left(\mathbf{x}_+^{(k)} \right)$ the use of an initial value algorithm that computes switching points is sometimes more efficient, since the initial point $\mathbf{x}_+^{(k)}$ is known. Hence, the time $t^{(k+1)}$ where the function g_i changes the sign can be computed very fast by using an algorithm to find zeros of a non-linear function.

$$
\begin{aligned}
\begin{bmatrix} \mathbf{x} \\ t^{(k+1)} \end{bmatrix}' &= \begin{bmatrix} t^{(k+1)}\mathbf{f}_i(\mathbf{x},\mathbf{p}) \\ 0 \end{bmatrix} \\
\mathbf{r}(\mathbf{x}(0)\,,\,\mathbf{x}(1)) &= \begin{bmatrix} \mathbf{x}(0) - \mathbf{x}_+^{(k)} \\ g_i(\mathbf{x}(1)) \end{bmatrix} = 0\,. \\
\mathbf{x}_-^{(k+1)} &:= \mathbf{x}(1) \\
\Psi_i\left(\mathbf{x}_+^{(k)}\right) &:= \sigma_i\left(\mathbf{x}_-^{(k+1)}\right)
\end{aligned}
\tag{3.8}
$$

Whether the trajectory Φ_ξ stays in X_i or intersects $g^{-1}(0)$ transversally at

$$\mathbf{x}_-^{(k+1)} := \mathbf{x}(1)$$

can be decided by the direction of Φ_ξ at point $\mathbf{x}_-^{(k+1)}$:

$$
\begin{aligned}
\mathbf{f}_i\left(\mathbf{x}_-^{(k+1)},\mathbf{p}\right) &\in T_{\mathbf{x}_-^{(k+1)}} g_i^{-1}(0) &\Rightarrow&\quad \Phi_\xi \text{ stays in } X_i\,, \\
\mathbf{f}_i\left(\mathbf{x}_-^{(k+1)},\mathbf{p}\right) &\notin T_{\mathbf{x}_-^{(k+1)}} g_i^{-1}(0) &\Rightarrow&\quad \Phi_\xi \text{ leaves } X_i\,.
\end{aligned}
\tag{3.9}
$$

$T_{\mathbf{x}_-^{(k+1)}} g_i^{-1}(0)$ is the tangent space of the manifold $g_i^{-1}(0)$ at $\mathbf{x}_-^{(k+1)}$ (cf. fig. 3.8). Therefore equation (3.9) is equivalent to

$$
\begin{aligned}
Dg\left(\mathbf{x}_-^{(k+1)}\right)^T \mathbf{f}_i\left(\mathbf{x}_-^{(k+1)},\mathbf{p}\right) &= 0 &\Rightarrow&\quad \Phi_\xi \text{ stays in } X_i\,, \\
Dg\left(\mathbf{x}_-^{(k+1)}\right)^T \mathbf{f}_i\left(\mathbf{x}_-^{(k+1)},\mathbf{p}\right) &\neq 0 &\Rightarrow&\quad \Phi_\xi \text{ leaves } X_i\,.
\end{aligned}
\tag{3.10}
$$

In some cases the solution Φ_ξ on X_i can be expressed analytically. Then the boundary value problem according to equation (3.8) is only a non linear algebraic equation. A practical way to compute the sequence $\left(\mathbf{x}_+^{(i)}\right)_{i\in\mathbb{N}}$ is to start with $\boldsymbol{\xi} := \mathbf{x}_+^{(k)}$ and obtain $\mathbf{x}_+^{(k+1)}$ from

$$\begin{aligned} & \left(g_i \circ \Phi_{\mathbf{x}_+^{(k)}}\right)(t^{(k+1)}) = 0 \Rightarrow t^{(k+1)} \quad [13] \\ \Rightarrow \quad & \mathbf{x}_+^{(k+1)} = \Phi_{\mathbf{x}_+^{(k)}}\left(t^{(k+1)}\right) \quad . \end{aligned} \tag{3.11}$$

In most systems – especially in systems with impacts – the zeroset $g^{-1}(0)$ is not a connected component. In that case it has to be additionally decided what connected component is intersected by the trajectory $\Phi_{\mathbf{x}^{(k)}}$.

To describe the situation mathematically, we assume that $g_i^{-1}(0)$ consists of $N \in \mathbb{N}$ connected components Z_{ij}, $(j = 1, \ldots, N)$:

$$g_i^{-1}(0) = \bigcup_{j=1}^{N} Z_{ij} \ . \tag{3.12}$$

Furthermore, we assume that Φ_ξ has entered X_i at $\mathbf{x}^{(k)} \in g_i^{-1}(0)$ and Φ_ξ has left X_i again on $g_i^{-1}(0)$ (cf. fig. 3.8). The vector field $\mathbf{f}_i$ was presumed to be LIPSCHITZ-continuous. Therefore – if Φ_ξ leaves $g_i^{-1}(0)$ – there is exactly one $\mathbf{x}^{(k+1)} \in Z_{ij}$ for an appropriate $j \in \{1, \ldots, N\}$ to each $\mathbf{x}^{(k)} \in g_i^{-1}(0)$. Hence, there is a unique "indicator function"

$$h_i : g_i^{-1}(0) \to \{1, \ldots, N\} \ , \quad \mathbf{x}^{(k)} \mapsto j \tag{3.13}$$

that maps any entry point $\mathbf{x}^{(k)}$ into that connected component Z_j or index j respectively where the trajectory $\Phi_{\mathbf{x}^{(k)}}$ went through. The knowledge of h_i makes the computation of the state $\mathbf{x}^{(k+1)}$ mostly easier. On the other hand, the numerical computation of h_i is very time-consuming. Therefore it is appropriate to decide in each single case, whether it is easier to compute h_i or to investigate the boundary value problem according to equation (3.8) or the non-linear system of equation (3.11) with respect to its solvability for each connected component.

∇

Example 3.3: (one-staged rattling model)

The analytical solution after the k-th impact (state $\mathbf{x}^{(k)}$, set X_i) for the model used in example 3.1 is:

[13] For the sake of clarity the "+" sign at the state $\mathbf{x}_+^{(k)}$ is ignored.

$$\Phi_{\mathbf{x}^{(k)}}(t) = \begin{bmatrix} \varphi_1\left(t, \mathbf{x}^{(k)}\right) \\ \varphi_2\left(t, \mathbf{x}^{(k)}\right) \\ \varphi_3\left(t, \mathbf{x}^{(k)}\right) \end{bmatrix}$$

$$= \begin{bmatrix} x_1^{(k)} + \left(\frac{m}{\delta}\right)\left(x_2^{(k)} - \dot{x}_p\left(x_3^{(k)}\right)\right)\left[1 - e^{-\frac{\delta}{m}\left(t - x_3^{(k)}\right)}\right] + \left(x_p(t) - x_p\left(x_3^{(k)}\right)\right) \\ \left(x_2^{(k)} - \dot{x}_p\left(x_3^{(k)}\right)\right) e^{-\frac{\delta}{m}\left(t - x_3^{(k)}\right)} + \dot{x}_p(t) \\ t \end{bmatrix} .$$

x_p is a particular solution depending on the excitation function e. For

$$e(t) := A \sin \omega t$$

the particular solution is

$$x_p(t) = \frac{A\omega}{\left(\frac{\delta}{m}\right)^2 + \omega^2} \left[-\left(\frac{\delta}{m}\right) \cos \omega t + \omega \sin \omega t\right] .$$

The state space TM is bounded with respect to coordinate x_1. The only possibility for an impact is at $x_1 = 0$ or $x_1 = v$. The time $x_3^{(k+1)}$ where the $(k+1)$-th impact takes place and the velocity $x_2^{(k+1)}$ immediately before impact are not bounded. The simplest limit function that carries this property is

$$g(\mathbf{x}) := g_i(\mathbf{x}) := x_1(x_1 - v) \quad ^{14} .$$

The zeroset $g^{-1}(0)$ is not connected. However its connected components

$$Z_0 := \left\{(0, x_2, x_3) \mid (x_2, x_3) \in \mathbb{R}^2\right\} ,$$

$$Z_v := \left\{(v, x_2, x_3) \mid (x_2, x_3) \in \mathbb{R}^2\right\}$$

are $dim(TM) - 1 = 2$ dimensional submanifolds ($g^{-1}(0) = Z_0 \cup Z_v$).

[14] The limit function g_i is the same on each subset X_i. Therefore the index i will be ignored in the following.

According to equation (3.11) the time $t^{(k+1)} := x_3^{(k+1)}$ ($(k+1)$-th impact) can be computed by the equation

$$\begin{aligned} 0 &= \left(g \circ \Phi_{\mathbf{x}^{(k)}}\right)\left(t^{(k+1)}\right) , \\ &= \varphi_1\left(\mathbf{x}^{(k)}, t^{(k+1)}\right)\left[\varphi_1\left(\mathbf{x}^{(k)}, t^{(k+1)}\right) - v\right] . \end{aligned}$$

The equation holds, iff

$$\varphi_1\left(\mathbf{x}^{(k)}, t^{(k+1)}\right) = 0$$

or

$$\varphi_1\left(\mathbf{x}^{(k)}, t^{(k+1)}\right) = v$$

is the case.

Only one of these equations has a solution. The relevant equation can be determined by the indicator function h_i according to equation (3.13).

In systems with impact and especially in the model with one degree of freedom the numerical computation of h_i is simple:

$\mathbf{x}^{(k)} \in Z_0$, implies $\mathbf{x}^{(k+1)} \in Z_v$ or $\mathbf{x}^{(k+1)} \in Z_0$ (cf. fig. 3.9).

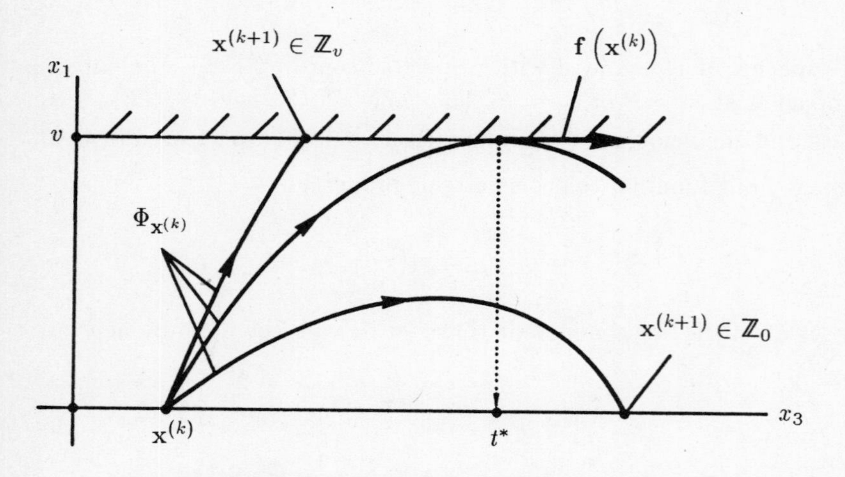

Fig. 3.9: determination of the indicator function h

The limit case occurs if trajectory $\Phi_{\mathbf{x}^{(k)}}$ touches the separatrix tangentially, that is if $f\left(\mathbf{x}^{(k+1)}\right) \in Z_v$ holds.

Or more precise, if

$$\begin{aligned} \varphi_1\left(\mathbf{x}^{(k)}, t^*\right) &= v \;, \\ \varphi_2\left(\mathbf{x}^{(k)}, t^*\right) &= 0 \end{aligned} \qquad (A)$$

holds.

In equation (A) time t^* is an additional unknown parameter. t^* can be eliminated, but usually it is no problem to determine t^*. For the following considerations t^* is not of importance.

If N_0 is the zeroset of equation (A), the area of separatrix Z_0 is separated by N_0 into two disjunct open subsets U_v and U_0, namely into all states $\mathbf{x}^{(k)}$ which lead to a state $\mathbf{x}^{(k+1)} \in Z_v$ and into all states $\mathbf{x}^{(k)}$ which lead to a state $\mathbf{x}^{(k+1)} \in Z_0$.

Hence, the indicator function h_i can be expressed by

$$h_i\left(\mathbf{x}^{(k)}\right) = \begin{cases} 0 & \text{if} \quad \mathbf{x}^{(k)} \in U_0 \\ v & \text{if} \quad \mathbf{x}^{(k)} \in U_v \;. \end{cases}$$

An analogous indicator function is obtained for states $\mathbf{x}^{(k)} \in Z_v$. In that case, the equations

$$\begin{aligned} \varphi_1\left(\mathbf{x}^{(k)}, t^*\right) &= 0 \;, \\ \varphi_2\left(\mathbf{x}^{(k)}, t^*\right) &= 0 \end{aligned} \qquad (B)$$

have to be solved.

The zeroset N_v of equation (B) separates Z_v into open disjunct subsets V_0 and V_v. Therefore the indicator function is given by:

$$h_i\left(\mathbf{x}^{(k)}\right) = \begin{cases} 0 & \text{if} \quad \mathbf{x}^{(k)} \in U_0 \cup V_0 \\ v & \text{if} \quad \mathbf{x}^{(k)} \in U_v \cup V_v \end{cases} \;.$$

For $e(t) = A \sin \omega t$ the numerically computed domains U_0 and U_v or V_0 and V_v respectively are shown in fig. 3.10.

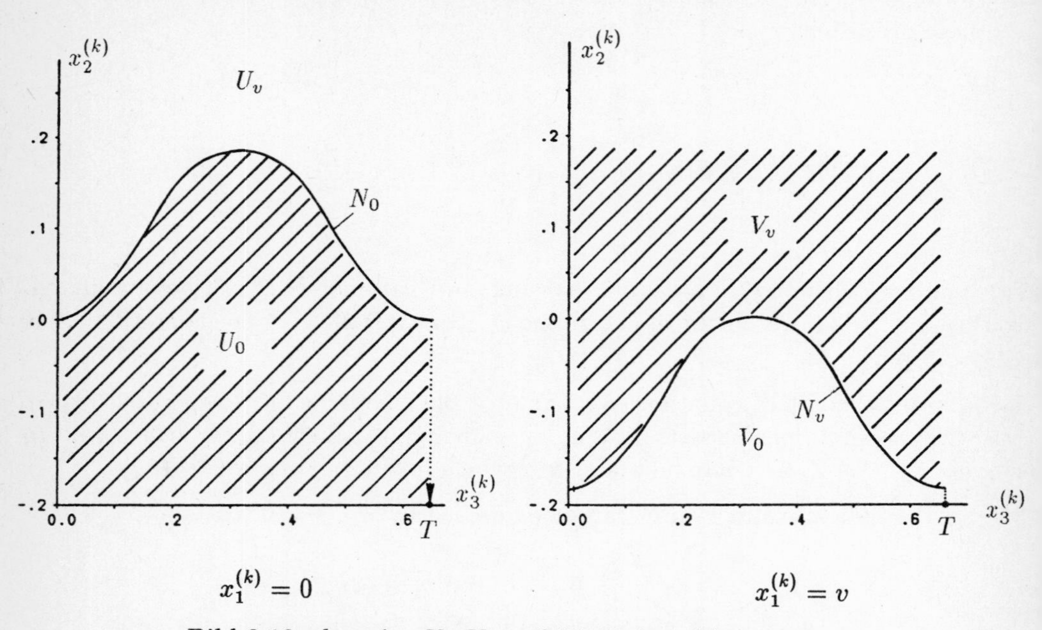

Bild 3.10: domains U_0, U_v and V_0, V_v for $e(t) = A \sin \omega t$

Function e is $T = \frac{2\pi}{\omega}$ periodic with respect to time t. Hence, domains of coordinate $x_3^{(k)}$ are T-periodic.

The recursion $\boldsymbol{\Psi}_i$ is given by

$$\boldsymbol{\Psi}_i\left(\mathbf{x}^{(k)}\right) := \begin{bmatrix} \left|x_1^{(k)} - \frac{v}{2}\right| \left(sgn\left[x_2^{(k)} - \gamma\left(x_1^{(k)},\, x_3^{(k)}\right)\right] + 1\right) \\ -\varepsilon\, \varphi_2\left(x_3^{(k+1)},\, \mathbf{x}^{(k)}\right) \\ \varphi_1^{-1}\left(x_3^{(k+1)},\, \mathbf{x}^{(k)}\right) \end{bmatrix}$$

where $\gamma : \{0, v\} \times \mathbb{R} \to \mathbb{R}$, $\left(x_1^{(k)}, x_3^{(k)}\right) \mapsto x_2^{(k)}$ is the limit function that describes the zerosets N_0 and N_v and

$$\sigma_i\left(\mathbf{x}^{(k)}\right) := \begin{bmatrix} 1 & 0 & 0 \\ 0 & -\varepsilon & 0 \\ 0 & 0 & 1 \end{bmatrix} \mathbf{x}^{(k)}$$

is the "transfer function" according to equation (3.6).

$\triangle$

∇

Example 3.4: (dry friction model)

Also for the model with one degree of freedom described in example 3.2 an analytical solution on X_1 and X_2 exists. For $\mathbf{x}^{(k)} \in G$ (cf. fig. 3.5) and small damping ($\delta < 2\sqrt{mc}$) the solution is given by

$$\Phi_{\mathbf{x}^{(k)}}(t) = \begin{bmatrix} \varphi_1\left(t, \mathbf{x}^{(k)}\right) \\ \varphi_2\left(t, \mathbf{x}^{(k)}\right) \\ \varphi_3\left(t, \mathbf{x}^{(k)}\right) \end{bmatrix}$$

$$= \begin{bmatrix} e^{-dt}\left[a(\mathbf{x}^{(k)})\cos(\Omega t) + b(\mathbf{x}^{(k)})\sin\Omega t\right] + x_p(t) \mp \frac{\mu g}{\nu^2} \\ e^{-dt}\left[\left(b(\mathbf{x}^{(k)})\Omega - da(\mathbf{x}^{(k)})\right)\cos\Omega t + \left(-a(\mathbf{x}^{(k)})\Omega - db(\mathbf{x}^{(k)})\right)\sin\Omega t\right] + \dot{x}_p(t) \\ t \end{bmatrix}$$

with $\Omega^2 := \nu^2 - d^2$; $d := \frac{\delta}{2m}$; $\nu^2 := \frac{c}{m}$.

$a\left(\mathbf{x}^{(k)}\right)$ and $b\left(\mathbf{x}^{(k)}\right)$ are constant, computed from equation $\Phi_{\mathbf{x}^{(k)}}\left(x_3^{(k)}\right) = \mathbf{x}^{(k)}$.

The "+" sign in the definition of $\Phi_{\mathbf{x}^{(k)}}$ is relevant if $\Phi_{\mathbf{x}^{(k)}}$ enters X_i, otherwise the "−" sign holds.

$x_p : \mathbb{R} \to \mathbb{R}$ is a particular solution with respect to function e. For $e(t) := A \cdot \sin\omega t$ the particular solution is

$$x_p(t) = \left(\frac{A\omega^2}{\left(\nu^2 - \omega^2\right)^2 + 4d^2\omega^2}\right)\left(\left(\nu^2 - \omega^2\right)\sin(\omega t) - 2d\omega\cos(\omega t)\right) .$$

The limit function g, whose zeroset separates TM into X_1 and X_2, can be easily described by

$$g(\mathbf{x}) := x_2$$

$$\Rightarrow \quad g^{-1}(0) = \{\mathbf{x} \in TM \mid x_2 = 0\} = G \cup H .$$

$g^{-1}(0)$ is connected. Therefore it is not necessary to determine the indicator function h according to equation (3.13).

$t^{(k+1)} := x_3^{(k+1)}$ can be obtained from (cf. equation (3.11))

$$0 = \left(g \circ \Phi_{\mathbf{x}^{(k)}}\right)\left(t^{(k+1)}\right) ,$$

$$= \varphi_2\left(t^{(k+1)}, \mathbf{x}^{(k)}\right) .$$

Transfer function σ_i is the identity:

$$\sigma_i\left(\mathbf{x}^{(k)}\right) = \mathbf{x}^{(k)} \ .$$

Hence, the recursion function $\boldsymbol{\Psi}_i$ at $\mathbf{x}^{(k)} \in G$ is:

$$\boldsymbol{\Psi}_i\left(\mathbf{x}^{(k)}\right) := \begin{bmatrix} \varphi_1\left(x_3^{(k+1)},\, \mathbf{x}^{(k)}\right) \\ 0 \\ \varphi_2^{-1}\left(0,\, \mathbf{x}^{(k)}\right) \end{bmatrix} .$$

$i = 1$ if $\Phi_{\mathbf{x}^{(k)}}$ enters X_1 and $i = 2$ if $\Phi_{\mathbf{x}^{(k)}}$ enters X_2. There are two possibilities for $\mathbf{x}^{(k)} \in H$:

Either the trajectory $\Phi_{\mathbf{x}^{(k)}}(t)$ stays in H for all times $t > x_3^{(k)}$ or $\Phi_{\mathbf{x}^{(k)}}(t)$ leaves the sticking domain H again. If $\Phi_{\mathbf{x}^{(k)}}$ leaves H, then it does so at point

$$\mathbf{x}^+ := \left(x_1^{(k)}\,,\, x_2^{(k)}\,,\, t^+\right)^T$$

or at point

$$\mathbf{x}^- := \left(x_1^{(k)}\,,\, x_2^{(k)}\,,\, t^-\right)^T \ .$$

t^+ or t^- respectively has to be determined from the non-linear algebraic equation

$$r^{\overset{+}{(-)}}\left(t, \mathbf{x}^{(k)}, \mathbf{p}\right) := \left(\frac{c}{m}\right) x_1^{(k)} + \ddot{e}(t, p_i) \ \underset{(+)}{-}\ \mu g = 0 \ .$$

To decide a priori whether sticking or slipping occurs at $\mathbf{x}^-$ or at $\mathbf{x}^+$, it is useful to divide H into subsets H^+ and H^- defined by

$$H^{\overset{+}{(-)}} := \left\{\mathbf{x} \in H \ | \ \exists\, t^* > x_3 \text{ so that } r^{\overset{+}{(-)}}(t^*, \mathbf{x}, \mathbf{p}) \le 0 \text{ and } \{x_1\} \times [x_3, t^*] \subset H\right\} .$$

Fig. 3.11 shows the partition for $e(t) = A \sin \omega t$.

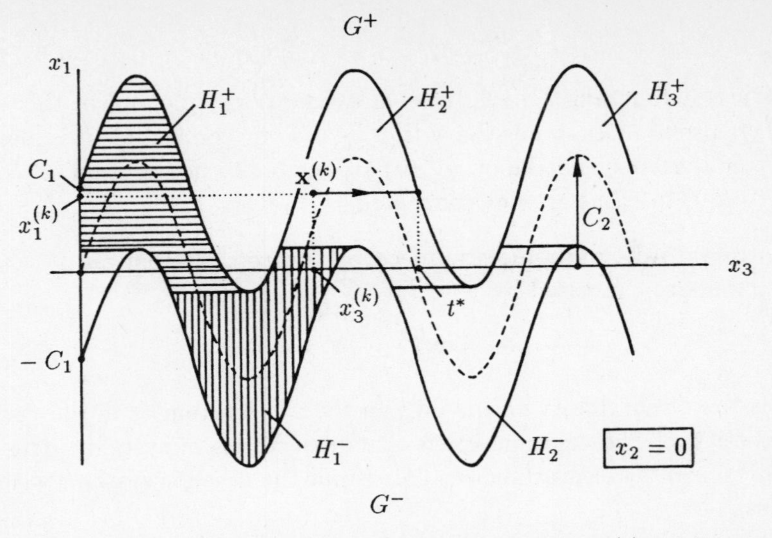

Fig. 3.11: partition of the stick domain H for $e(t) = A \sin \omega t$

In that case $H^{\overset{+}{(-)}}$ is defined by

$$H^{\overset{+}{(-)}} = \bigcup_{j \in \mathbb{Z}} H_j^{\overset{+}{(-)}}$$

with

$$C_1 := \left(\frac{\mu g m}{c}\right) \quad ; \qquad C_2 := \left(\frac{A\omega^2 m}{c}\right) .$$

The excitation function is $T = \frac{2\pi}{\omega}$-periodic, hence the domains $H_j^{\overset{+}{(-)}}$ are also periodically repeated.

Analogously to the rattling model (example 3.3) the indicator function assigns a $\mathbf{x}^{(k)} \in H$ to the domain G^+ (G^-) if the trajectory $\Phi_{\mathbf{x}^{(k)}}$ enters X_1 (X_2) (cf. fig. 3.5). That means

$$h\left(\mathbf{x}^{(k)}\right) := \begin{cases} + & \text{if} \quad \mathbf{x}^{(k)} \in H^+ \\ - & \text{if} \quad \mathbf{x}^{(k)} \in H^- \end{cases} .$$

$h\left(\mathbf{x}^{(k)}\right) = +\,(-)$ says that the trajectory enters at $\mathbf{x}^+$ $(\mathbf{x}^-)$. Immediately after the entrance, recursion $\boldsymbol{\Psi}_i$ holds.

If χ^+ (χ^-) is the function that maps $\mathbf{x}^{(k)}$ into $\mathbf{x}^+(\mathbf{x}^-)$, the recursion can be expressed by

$$\mathbf{x}^{(k+1)} = \begin{cases} \boldsymbol{\Psi}_i\left(\mathbf{x}^{(k)}\right) & \text{if} \quad \mathbf{x}^{(k)} \in G \\ (\boldsymbol{\Psi}_i \circ \chi^-)(\mathbf{x}^{(k)}) & \text{if} \quad \mathbf{x}^{(k)} \in H^- \\ (\boldsymbol{\Psi}_i \circ \chi^+)(\mathbf{x}^{(k)}) & \text{if} \quad \mathbf{x}^{(k)} \in H^+ \end{cases} .$$

Remarks:

(1) An indicator function h, as defined in the last example and in this example, is always useful or even necessary if $g^{-1}(0)$ is not connected. h is also needed if different cases considered on $g^{-1}(0)$ are related to domains on $g^{-1}(0)$ – such as G^- and G^+ – that are not connected.

(2) In contrary to discontinuities induced by impacts, discontinuities caused by dry friction are expressed by

$$\mathbf{x}_-^{(k)} = \mathbf{x}_+^{(k)} \qquad \forall\ k \in \mathbb{N}\ .$$

That is the discontinuity occurs only in the acceleration or in the vector field $\mathbf{f}_i$ respectively. The consequence is that trajectories of systems including dry friction elements are continuous. This is not the case for systems with impact elements.

(3) For $c \to 0$ the domains G and H in case of $e(t) = A \sin \omega t$ change periodically (cf. fig. 3.12).

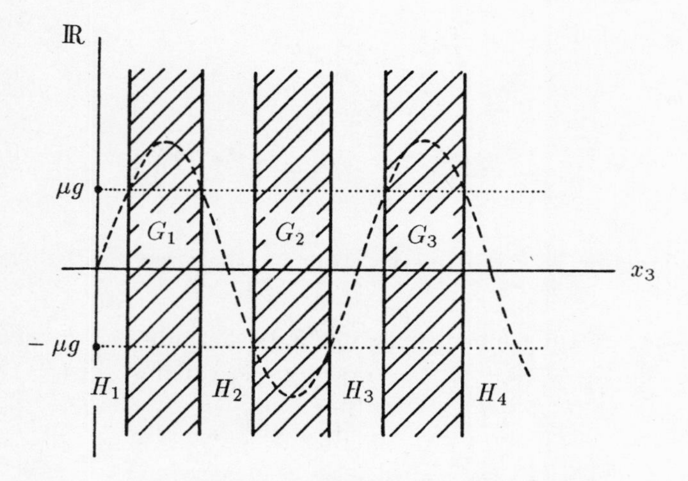

Fig. 3.12: stick/slip domains for $c = 0$

$$G = \bigcup_{i \in \mathbb{Z}} G_i \quad ; \qquad H = \bigcup_{i \in \mathbb{Z}} H_i \quad .$$

(4) For $e(t) = A \sin \omega t$ permanent sticking is only possible if the displacement C_1 is greater than the amplitude C_2, that is if

$$\frac{\mu g}{A\omega^2} > 1\ .$$

△

3.2 Periodic solutions of recursively described systems

As shown in chapter 3.1 the use of POINCARÉ sections $\Psi : g^{-1}(0) \to g^{-1}(0)$ is a proper way to describe a dynamical system recursively. In that case, the POINCARÉ section $g^{-1}(0)$ determines a $dim(TM)-1$ dimensional submanifold in the state space TM. And the discontinuities of the vector field are located on these manifolds.

3.2.1 Definition of a periodic solution

A unique piece of the trajectory $\Phi_{\mathbf{x}^{(k)}}$ exists between any two neighboring points $\mathbf{x}^{(k)}$, $\mathbf{x}^{(k+1)} \in g^{-1}(0)$ of the recursion Ψ. Hence, for each $\mathbf{z} \in g^{-1}(0)$ there is a unique recursion sequence

$$\Gamma_{\mathbf{z}} : \mathbb{N} \to g^{-1}(0) \tag{3.14}$$

which is defined by

$$\begin{aligned} \mathbf{x}^{(0)} &:= \mathbf{z} \ , \\ \mathbf{x}^{(k+1)} &:= \boldsymbol{\Psi}_{i_k}\left(\mathbf{x}^{(k)}\right) \ ; \quad k = 0, 1, 2, \ldots \ . \end{aligned} \tag{3.15}$$

The index i_k denotes the set X_{i_k} where the trajectory from $\mathbf{x}^{(k)}$ to $\mathbf{x}^{(k+1)}$ is embedded. Analogous to differentiable dynamical systems, it is useful to define a periodic solution as follows:

Definition (3.2):
A sequence $\Gamma_{\mathbf{z}} : \mathbb{N} \to g^{-1}(0)$ of recursion points is called periodic if there is a $m \in \mathbb{N}$ and – in non-autonomous cases – a $n \in \mathbb{N}$ so that

$$\Gamma_{\mathbf{z}}(m) = \Gamma_{\mathbf{z}}(0) + n\,T\,\mathbf{e}_{2f+1} \tag{3.16}$$

holds, i.e. if a $(m, n) \in \mathbb{N}^2$ exists so that

$$\mathbf{x}^{(m)} = \mathbf{z} + n\,T\,\mathbf{e}_{2f+1} \tag{3.17}$$

or

$$(\boldsymbol{\Psi}_{i_m} \circ \ldots \circ \boldsymbol{\Psi}_{i_1})(\mathbf{z}) = \mathbf{z} + n\,T\,\mathbf{e}_{2f+1} \tag{3.18}$$

respectively is true.

○

In that case, the trajectory $\Phi_{\mathbf{x}^{(k)}}$ intersects the sets $X_{i_1}, \ldots, X_{i_m}$. In the non-autonomous case, the time is defined by the coordinate x_{2f+1}. The unit vector $\mathbf{e}_{2f+1}$ says that the trajectory has the period $n \cdot T$. T is the minimal time of period of the vector field $\mathbf{f}$. In non-autonomous vector fields – such as in differentiable systems – $\mathbf{f}$ has to be periodic with respect to time t.

Now, in order to compute a periodic solution the numbers m and n and a vector $\boldsymbol{\tau} := (i_1, \ldots, i_m)$ which indicates the sets X_{i_k} sequentially passed by the periodic solution (cf. chapter 3.2.3), must be known.

Therefore, it is reasonable to introduce the following abbreviations:

Definition (3.3):

According to definition (3.1) a periodic solution is called a "$(m, n, \boldsymbol{\tau})$–cycle" or a "$(m : n)$–cycle" of the recursion $\boldsymbol{\Psi} : g^{-1}(0) \to g^{-1}(0)$.

○

3.2.2 Invariance of translation of a periodic solution

If $\Gamma_{\mathbf{z}} : \mathbb{N} \to g^{-1}(0)$ is a $(m : n)$–cycle of the recursion $\boldsymbol{\Psi} : g^{-1}(0) \to g^{-1}(0)$, then the periodicity is expressed by:

$$\Gamma_{\mathbf{z}}(m + k) = \Gamma_{\mathbf{z}}(k) + n\,T\,\mathbf{e}_{2f+1} \qquad \forall\ k \in \mathbb{N}\ . \tag{3.19}$$

That means the boundary value problem or fixed point problem respectively – i.e. find a $\mathbf{z} \in g^{-1}(0)$ so that the non-linear system of equations

$$\mathbf{r}\,(\Gamma_{\mathbf{z}}(0),\ \Gamma_{\mathbf{z}}(m),\ n) := \Gamma_{\mathbf{z}}(m) - \Gamma_{\mathbf{z}}(0) - n\,T\,\mathbf{e}_{2f+1} = 0 \tag{3.20}$$

has a solution – is not unique.

However, the ambiguousness does not lead – as it does in systems without discontinuities – to difficulties in computing $\mathbf{z}$ or $\Gamma_{\mathbf{z}}$ respectively. The reason is that all periodic solutions which belong to a given $(m, n, \boldsymbol{\tau})$ are locally isolated.

3.2.3 Numerical computation of periodic solutions

The computation of a periodic solution according to definition (3.2) can be reduced to the numerical computation of a solution $\mathbf{z}$ of the non-linear algebraic equation

$$\boldsymbol{\rho}(\mathbf{z}) := \mathbf{r}\,(\Gamma_{\mathbf{z}}(0),\ \Gamma_{\mathbf{z}}(m),\ n) = 0 \tag{3.21}$$

for a given m, n and $\boldsymbol{\tau}$.

The function $\boldsymbol{\rho}$ at point $\mathbf{z}$ is computed by a series of m boundary value problems according to equation (3.8) or by a successive computation of m initial value problems with switching points.

Algorithms to solve equation (3.21) require the JACOBIan matrix

$$D\boldsymbol{\rho}(\mathbf{z}) = D\boldsymbol{\Psi}_{i_m}\left(\mathbf{x}^{(m-1)}\right) \cdot D\boldsymbol{\Psi}_{i_{m-1}}\left(\mathbf{x}^{(m-2)}\right) \cdot \ldots \cdot D\boldsymbol{\Psi}_{i_1}(z) \ . \tag{3.22}$$

However, the computation of $D\boldsymbol{\rho}(\mathbf{z})$ is very time-consuming. Therefore it is more efficient to consider the points $\mathbf{x}^{(i)}$ as unknown – such as in the multiple shooting method – and compute the vector

$$\boldsymbol{\zeta} := \left(\mathbf{z}\,,\ \mathbf{x}^{(1)}\,,\ \ldots\,,\ \mathbf{x}^{(m-1)}\right) \tag{3.23}$$

from the modified system of equations

$$\mathbf{R}(\boldsymbol{\zeta}) := \begin{bmatrix} \boldsymbol{\Psi}_{i_1}(\mathbf{z}) - \mathbf{x}^{(1)} \\ \vdots \\ \boldsymbol{\Psi}_{i_{m-1}}\left(\mathbf{x}^{(m-2)}\right) - \mathbf{x}^{(m-1)} \\ \boldsymbol{\Psi}_{i_m}\left(\mathbf{x}^{(m-1)}\right) - \mathbf{z} - n \cdot T\,\mathbf{e}_{2f+1} \end{bmatrix} = \mathbf{0}\ . \tag{3.24}$$

The start of the iteration requires more estimations for the initial value, but the necessary decomposition of the JACOBIan matrix

$$D\mathbf{R}(\boldsymbol{\zeta}) = \begin{bmatrix} \mathbf{G}_1 & -\mathbf{E} & & \mathbf{0} \\ & \ddots & \ddots & \\ & \mathbf{0} & \ddots & -\mathbf{E} \\ -\mathbf{E} & & & \mathbf{G}_m \end{bmatrix} \tag{3.25}$$

with

$$\mathbf{G}_k := D\boldsymbol{\Psi}_{i_k}\left(\mathbf{x}^{(k-1)}\right) \qquad (k = 1, \ldots, m) \tag{3.26}$$

can be done by less numerical operations.

The reason is – such as in the multiple shooting method – that by using the unit matrices $\mathbf{E}$ on the superdiagonal as PIVOT–elements the decomposition of $D\mathbf{R}(\boldsymbol{\zeta})$ requires only $(m-1)$ block-GAUSS steps.

∇

Example 3.5: (rattling model with 1 degree of freedom)

For the one-staged rattling model according to example 3.1 a (2 : 1) cycle will be considered. The excitation function is given by $e(t) = A \sin \omega t$. The vector $\boldsymbol{\tau}$ is not necessary since the sets X_i repeat in a T-periodic sequence $\left(T = \frac{2\pi}{\omega}\right)$.

Furthermore, in that case it can be assumed – without any restriction – that $z_1 = x_1^{(0)} = 0$ and $x_1^{(1)} = v$ holds. Based on this assumption, the recursion is given by

$$\boldsymbol{\Psi}_1(\mathbf{z}) \quad = \quad \begin{bmatrix} v \\ -\varepsilon\, \varphi_2\left(x_3^{(1)}\,,\, \mathbf{z}\right) \\ \varphi_1^{-1}\left(x_1^{(1)}\,,\, \mathbf{z}\right) \end{bmatrix} \quad \in \mathbb{R}^3\,,$$

$$\boldsymbol{\Psi}_2(\mathbf{x}^{(1)}) \quad = \quad \begin{bmatrix} 0 \\ -\varepsilon\, \varphi_2\left(z_3 + T\,,\, \mathbf{x}^{(1)}\right) \\ \varphi_1^{-1}\left(z_1^{(1)}\,,\, \mathbf{x}^{(1)}\right) \end{bmatrix} \quad \in \mathbb{R}^3\,,$$

and therefore the non-linear function $\mathbf{R}(\boldsymbol{\zeta})$ with $\boldsymbol{\zeta} := \left(\mathbf{z}, \mathbf{x}^{(1)}\right)$ is determined by:

$$\mathbf{R}(\boldsymbol{\zeta}) = \begin{bmatrix} \Psi_1(\mathbf{z}) - \mathbf{x}^{(1)} \\ \Psi_2(\mathbf{x}^{(1)}) - \mathbf{z} \end{bmatrix} \quad \in \mathbb{R}^6\,.$$

The JACOBIan matrix $D\mathbf{R}(\boldsymbol{\zeta})$ can be expressed by

$$D\mathbf{R}(\boldsymbol{\zeta}) = \left[\begin{array}{c|c} D\boldsymbol{\Psi}_1(\mathbf{z}) & -\mathbf{E} \\ \hline -\mathbf{E} & D\boldsymbol{\Psi}_2\left(\mathbf{x}^{(1)}\right) \end{array}\right] \quad \in \ \mathbb{R}^{6,6}$$

with

$$D\boldsymbol{\Psi}_1(\mathbf{z}) = \begin{bmatrix} 0 \\ -\varepsilon\left(\frac{\partial}{\partial \mathbf{z}}\left[\varphi_2\left(x_3^{(1)}, \mathbf{z}\right)\right] + \frac{\partial}{\partial x_3^{(1)}}\left[\varphi_2\left(x_3^{(1)}, \mathbf{z}\right)\right] \cdot \frac{\partial x_3^{(1)}}{\partial \mathbf{z}}\right) \\ \frac{\partial x_3^{(1)}}{\partial \mathbf{z}} \end{bmatrix} \quad \in \ \mathbb{R}^{3,3}$$

$$\left(\frac{\partial x_3^{(1)}}{\partial \mathbf{z}} := -\frac{\partial}{\partial \mathbf{z}}\left[\varphi_1\left(x_3^{(1)}, \mathbf{z}\right)\right] \ / \ \frac{\partial}{\partial x_3^{(1)}}\left[\varphi_1\left(x_3^{(1)}, \mathbf{z}\right)\right]\right)$$

and

$$D\boldsymbol{\Psi}_2(\mathbf{x}^{(1)}) \;=\; \begin{bmatrix} 0 \\ -\varepsilon\left(\frac{\partial}{\partial \mathbf{x}^{(1)}}\left[\varphi_2\left(z_3+T,\mathbf{x}^{(1)}\right)\right] + \frac{\partial}{\partial z_3}\left[\varphi_2\left(z_3+T,\mathbf{x}^{(1)}\right)\right]\cdot\frac{\partial z_3}{\partial \mathbf{x}^{(1)}}\right) \\ \frac{\partial z_3}{\partial \mathbf{x}^{(1)}} \end{bmatrix}$$

$$\in \mathbb{R}^{3,3} \;,$$

$$\left(\frac{\partial z_3}{\partial \mathbf{x}^{(1)}} := -\frac{\partial}{\partial \mathbf{x}^{(1)}}\left[\varphi_1\left(z_3+T,\mathbf{x}^{(1)}\right)\right] \;/\; \frac{\partial}{\partial z_3}\left[\varphi_1\left(z_3+T,\mathbf{z}\right)\right]\right) \;.$$

The functions φ_1 and φ_2 are the same as in example 3.3.

A solution of the non-linear equation $\mathbf{R}(\boldsymbol{\zeta}) = 0$ can be computed by standard algorithms of software libraries, which include an explicit input of the matrix $D\mathbf{R}(\boldsymbol{\zeta})$ or its triangular decomposition.

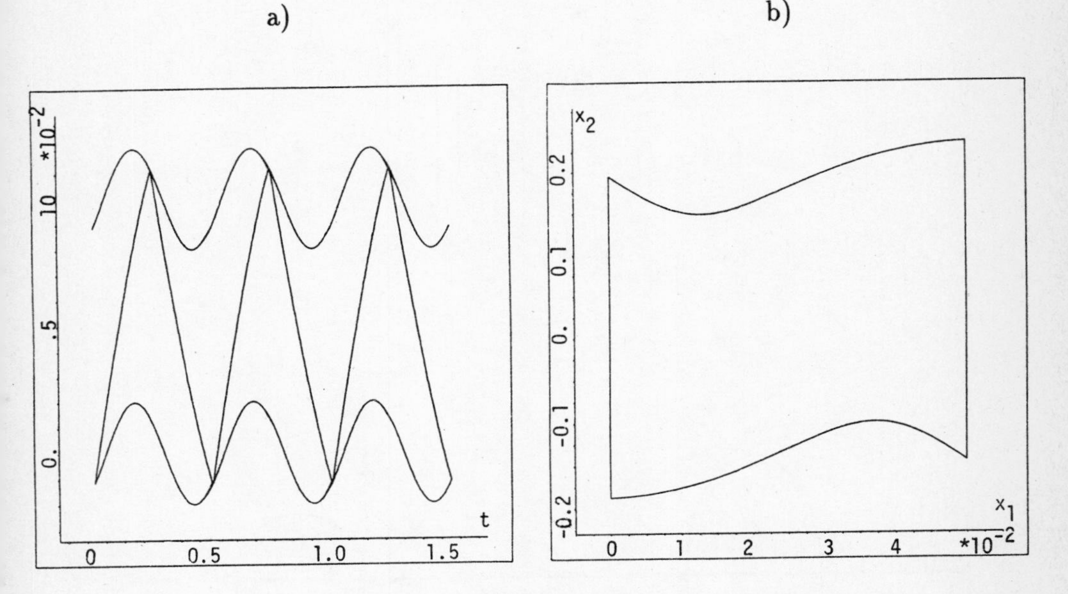

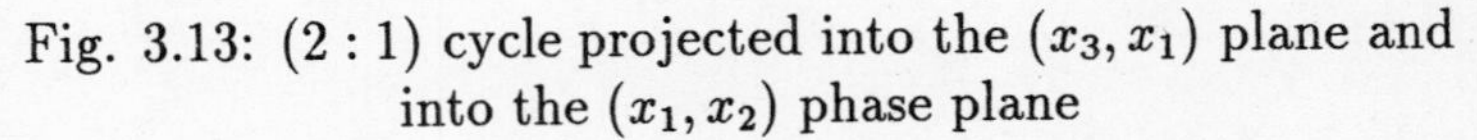

Fig. 3.13: (2 : 1) cycle projected into the (x_3, x_1) plane and into the (x_1, x_2) phase plane

Fig. 3.13 shows the time history of the (2 : 1) cycle with period $T = \frac{2\pi}{\omega}$. The values are related to the parameters given in the appendix.

$\triangle$

∇

Example 3.6: (dry friction model)

As in example 3.5, a (2 : 1) cycle for the dry friction model according to example 3.2 is considered. The excitation function is given by $e(t) = A \sin \omega t$. Damping δ is neglected ($\delta = 0$).

A trajectory on X_1 ("+" sign) or X_2 ("−" sign) has the following analytical expression:

$$\Phi_{\mathbf{x}^{(k)}}(t) = \begin{bmatrix} \varphi_1\left(t, \mathbf{x}^{(k)}\right) \\ \varphi_2\left(t, \mathbf{x}^{(k)}\right) \\ \varphi_3\left(t, \mathbf{x}^{(k)}\right) \end{bmatrix} = \begin{bmatrix} a \cos \Omega t + b \sin \Omega t + x_p(t) \underset{(+)}{-} \frac{\mu g}{\Omega^2} \\ \Omega(b \cos \Omega t - a \sin \Omega t) + \dot{x}_p(t) \\ t \end{bmatrix}$$

with

$$\Omega := \sqrt{\frac{c}{m}} ,$$

$$\begin{bmatrix} a \\ b \end{bmatrix} = \begin{bmatrix} \cos\left(\Omega x_3^{(k)}\right) & -\sin\left(\Omega x_3^{(k)}\right) \\ \sin\left(\Omega x_3^{(k)}\right) & \cos\left(\Omega x_3^{(k)}\right) \end{bmatrix} \begin{bmatrix} x_1^{(k)} \underset{(-)}{+} \frac{\mu g}{\Omega^2} - x_p\left(x_3^{(k)}\right) \\ \left(\frac{x_2^{(k)} - \dot{x}_p\left(x_3^{(k)}\right)}{\Omega}\right) \end{bmatrix} ,$$

$$x_p(t) = \left(\frac{A\omega^2}{\nu^2 - \omega^2}\right) \sin \omega t .$$

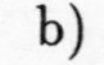

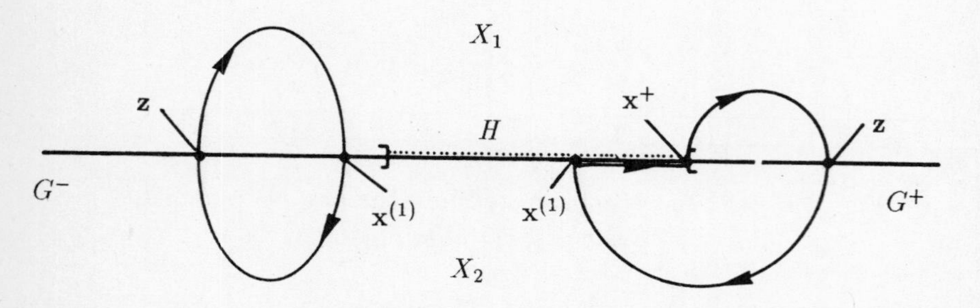

Fig. 3.14: qualitative behavior of a (2 : 1) cycle with and without sticking

The limit cycle either intersects the sticking area H or it does not touch H (cf. fig. 3.14). Whether the cycle intersects H or not depends on the initial value $\mathbf{z}$ and the parameter $\mathbf{p}$.

1st case: sticking does not occur (fig. 3.14 a)

$$\Rightarrow \qquad \boldsymbol{\Psi}_1(\mathbf{z}) = \begin{bmatrix} \varphi_1\left(x_3^{(1)}, \mathbf{z}\right) \\ 0 \\ \varphi_2^{-1}(0, \mathbf{z}) \end{bmatrix} \in \mathbb{R}^3 \qquad (\text{auf } X_1) \quad ,$$

$$\boldsymbol{\Psi}_2\left(\mathbf{x}^{(1)}\right) = \begin{bmatrix} \varphi_1\left(z_3 + T, \mathbf{x}^{(1)}\right) \\ 0 \\ \varphi_2^{-1}\left(0, \mathbf{x}^{(1)}\right) \end{bmatrix} \in \mathbb{R}^3 \qquad (\text{auf } X_2) \quad .$$

$\boldsymbol{\zeta}, \mathbf{R}(\boldsymbol{\zeta})$ and $D\mathbf{R}(\boldsymbol{\zeta})$ of example 3.5 and example 3.6 are identical.

$$D\boldsymbol{\Psi}_1(\mathbf{z}) = \begin{bmatrix} \frac{\partial}{\partial \mathbf{z}}\left[\varphi_1\left(x_3^{(1)}, \mathbf{z}\right)\right] + \frac{\partial}{\partial x_3^{(1)}}\left[\varphi_1\left(x_3^{(1)}, \mathbf{z}\right)\right] \cdot \frac{\partial x_3^{(1)}}{\partial \mathbf{z}} \\ \mathbf{0} \\ \frac{\partial x_3^{(1)}}{\partial \mathbf{z}} \end{bmatrix} \in \mathbb{R}^{3,3} \quad ,$$

$$\left(\frac{\partial x_3^{(1)}}{\partial \mathbf{z}} := -\frac{\partial}{\partial \mathbf{z}}\left[\varphi_2\left(x_3^{(1)}, \mathbf{z}\right)\right] \; / \; \frac{\partial}{\partial x_3^{(1)}}\left[\varphi_2\left(x_3^{(1)}, \mathbf{z}\right)\right] \right) \quad ,$$

$$D\boldsymbol{\Psi}_2(\mathbf{x}^{(1)}) = \begin{bmatrix} \frac{\partial}{\partial \mathbf{x}^{(1)}}\left[\varphi_1\left(z_3 + T, \mathbf{x}^{(1)}\right)\right] + \frac{\partial}{\partial z_3}\left[\varphi_1\left(z_3 + T, \mathbf{x}^{(1)}\right)\right] \cdot \frac{\partial z_3}{\partial \mathbf{x}^{(1)}} \\ \mathbf{0} \\ \frac{\partial z_3}{\partial \mathbf{x}^{(1)}} \end{bmatrix} \in \mathbb{R}^{3,3} \quad ,$$

$$\left(\frac{\partial z_3}{\partial \mathbf{x}^{(1)}} := -\frac{\partial}{\partial \mathbf{x}^{(1)}}\left[\varphi_2\left(z_3 + T, \mathbf{x}^{(1)}\right)\right] \; / \; \frac{\partial}{\partial z_3}\left[\varphi_2\left(z_3 + T, \mathbf{x}^{(1)}\right)\right] \right) \quad .$$

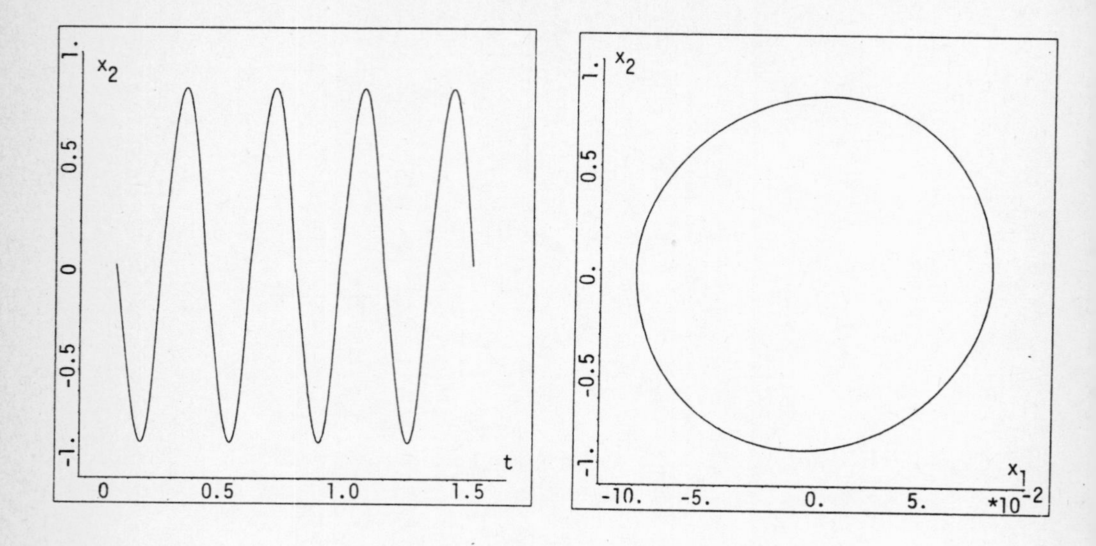

Fig. 3.15: (2 : 1) cycle without sticking $((x_2, x_3)$ plane and (x_1, x_2) plane)

Fig. 3.15 shows a (2 : 1) cycle without sticking projected into the (x_1, x_2) plane. The parameters A, ω, c, m and μ are listed in appendix.

<u>2nd case:</u> sticking occurs (fig. 3.14 b)

We want to assume that the cycle starts in the slipping area ($\mathbf{z} \in G$).

The functions $\boldsymbol{\Psi}_1(\mathbf{z})$ and $\boldsymbol{\Psi}_2(\mathbf{z})$ are the same like in case 1. $\mathbf{R}(\boldsymbol{\zeta})$ is given by

$$\mathbf{R}(\boldsymbol{\zeta}) := \begin{bmatrix} \boldsymbol{\Psi}_1(\mathbf{z}) - \mathbf{x}^{(1)} \\ \left(\boldsymbol{\Psi}_2 \circ \chi^{\binom{+}{-}}\right)\left(\mathbf{x}^{(1)}\right) - \mathbf{z} \end{bmatrix} \in \mathbb{R}^6 .$$

The "+" sign holds for $\mathbf{x}^{(1)} \in H^+$, the "−" sign holds for $\mathbf{x}^{(1)} \in H^-$. The JACOBIan $D\mathbf{R}(\boldsymbol{\zeta})$ can be expressed by

$$D\mathbf{R}(\boldsymbol{\zeta}) = \left[\begin{array}{c|c} D\boldsymbol{\Psi}_1(\mathbf{z}) & -\mathbf{E} \\ \hline -\mathbf{E} & D\left(\boldsymbol{\Psi}_2 \circ \mathbf{x}^{\binom{+}{-}}\right)\left(\mathbf{x}^{(1)}\right) \end{array}\right] \in \mathbb{R}^{6,6}$$

with

$$D\left(\boldsymbol{\Psi}_2 \circ \chi^{\binom{+}{-}}\right)\left(\mathbf{x}^{(1)}\right) = D\boldsymbol{\Psi}_2\left(\mathbf{x}^{\binom{+}{-}}\right) \cdot D\chi^{\binom{+}{-}}\left(\mathbf{x}^{(1)}\right)$$

and

$$D\chi^{(\overset{+}{-})}\left(\mathbf{x}^{(1)}\right) = \begin{bmatrix} 1 & 0 & 0 \\ 0 & 1 & 0 \\ -\frac{c}{m\frac{d^3e}{dt^3}(t^{(\overset{+}{-})},p_i)} & 0 & 0 \end{bmatrix}.$$

The JACOBIans $D\Psi_1(\mathbf{z})$ and $D\Psi_2(\mathbf{x}^{(\overset{+}{-})})$ are the same like in case 1 if $\mathbf{x}^{(1)}$ is replaced by $\mathbf{x}^{(\pm)}$.

Fig. 3.16 shows the behavior of the numerically computed (2 : 1) cycle. The values are related to the dataset listed in appendix.

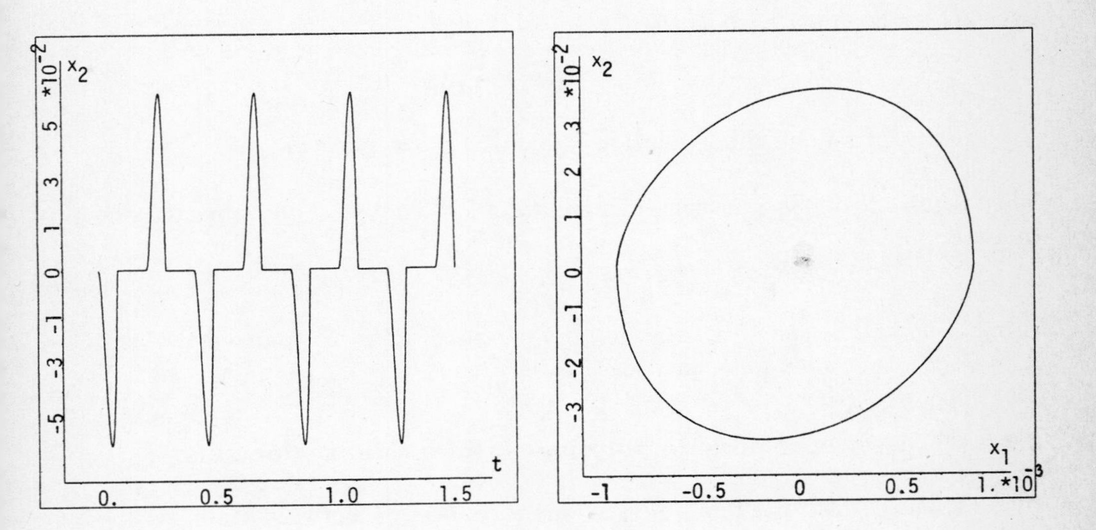

Fig. 3.16: limit cycle with sticking ((x_2, x_3) plane and (x_1, x_2) plane)

3.3 Stability and bifurcation of a sequence of recursively computed points

The main result in chapter 2.8 was that any asymptotically stable solution of a differentiable dynamical system has to converge to a limit cycle. Since any solution of the recursion defined in chapter 3.1 is unique, we have no difficulties to transfer the stability definition for differentiable systems to systems with discontinuities.

3.3.1 Definition of stability

The sequence $(\Gamma_{\mathbf{z}}(i))_{i\in\mathbb{N}}$ with $\mathbf{z} \in g^{-1}(0)$ defined by equation (3.15), is called stable if the following properties hold:

(a) $\Gamma_{\mathbf{z}} : \mathbb{N} \rightarrow g^{-1}(0) \ , \quad m \mapsto (\boldsymbol{\Psi}_{i_m} \circ \ldots \circ \boldsymbol{\Psi}_{i_1})(\mathbf{z})$ is bounded.

(b) There is an $\varepsilon(\mathbf{z}) > 0$ so that

$$\lim_{k\rightarrow\infty} \parallel \Gamma_{\mathbf{z}}(k) - \Gamma_{\mathbf{z}'}(k) \parallel = 0$$

holds for any neighbor solution $\Gamma_{\mathbf{z}'}$ with $\parallel \mathbf{z} - \mathbf{z}' \parallel < \varepsilon(\mathbf{z})$.

The magnitude of ε is a measure of stability of $\Gamma_{\mathbf{z}}$. The set of all connected domains of attraction

$$K(\mathbf{z}) := \left\{ \mathbf{z}' \in g^{-1}(0) \;\middle|\; \parallel \mathbf{z} - \mathbf{z}' \parallel < \varepsilon(\mathbf{z}) \right\} \tag{3.27}$$

is (cf. chapter 2.8) called the stability domain in Z^+. In the same way the stability domain Z^- is defined for sequences $(\Gamma_{\mathbf{z}}(-i))_{i\in\mathbb{N}}$.

3.3.2 Existence of periodic solutions inside stability domains

In accordance with differentiable systems a connection between stability domains and periodic solutions exist can be established by the following theorem:

<u>Theorem (3.1):</u>

For each stability domain exists exactly one periodic solution $\Gamma_{\mathbf{z}}$. Each sequence $(\Gamma_{\mathbf{z}'}(i))_{i\in\mathbb{N}}$ with $\mathbf{z}' \in Z^+$ converges to $\Gamma_{\mathbf{z}}$. That means, there is a $m \in \mathbb{N}$ and – for non-autonomous systems – a $n \in \mathbb{N}$ as well as a $\mathbf{z} \in Z^+$ so that $\Gamma_{\mathbf{z}}$ is a $(m : n)$ cycle.

<u>Proof:</u> as in theorem 2.4!

□

The same statement holds for domains Z^-. Hence, according to theorem (3.1) we have at least as many periodic solutions as stability domains exist.

3.3.3 Monodromy matrix of a $(m:n)$ cycle

Algorithms according to chapter 3.2 in order to compute limit cycles do not depend on the stability of the cycles. The decision about stability of periodic solutions can be made by knowledge about the distribution of the eigenvalues of the monodromy matrix shown in chapter 2.8.1.

To compute the monodromy matrix we assume that for a given $(m:n) \in \mathbb{N}^2$ the point $\mathbf{z}_0 \in g^{-1}(0)$ is a fixed point (periodic solution) of the mapping

$$\mathbf{F} : \mathbf{z} \mapsto \Gamma_{\mathbf{z}}(m) - n\,T\,\mathbf{e}_{2f} \ . \qquad (3.28)$$

That is $\Gamma_{\mathbf{z}_0}$ is a $(m:n)$ cycle.

In that case, a small disturbance $\mathbf{y} \in g^{-1}(0)$ is described by the mapping

$$\mathbf{y} \mapsto D\mathbf{F}(\mathbf{z}_0)\,\mathbf{y} \ . \qquad (3.29)$$

Hence, the monodromy matrix $D\mathbf{F}(\mathbf{z}_0)$ at $\mathbf{z}_0$ is

$$\begin{aligned} D\mathbf{F}(\mathbf{z}_0) &= D\Gamma_{\mathbf{z}_0}(m) \\ &= D\left(\boldsymbol{\Psi}_{i_m} \circ \ldots \circ \boldsymbol{\Psi}_{i_1}\right)(\mathbf{z}_0) \\ &= D\boldsymbol{\Psi}_{i_m}\left(\mathbf{z}^{(m-1)}\right) \cdot D\boldsymbol{\Psi}_{i_{m-2}}\left(\mathbf{z}^{(m-2)}\right) \cdot \ldots \cdot D\boldsymbol{\Psi}_1(\mathbf{z}_0) \\ &= \prod_{k=m}^{1} \mathbf{G}_{i_k} \qquad \text{(according to equation (3.26))} \ . \end{aligned} \qquad (3.30)$$

The JACOBIan matrices $\mathbf{G}_{i_k}$ are automatically produced by the computation of the zeros of equation (3.24). Therefore, the monodromy matrix of the computed periodic solution is already available.

According to equation (2.152) a necessary condition for stability of a $(m:n)$ cycle can be derived. The eigenvalues – except the one 1-eigenvalue – have to lie inside the unit circle of $\mathbb{C}$ (cf. equation (2.152)).

3.3.4 Bifurcation as a consequence of loss of stability

In differentiable systems with discontinuities the eigenvalues of the monodromy matrix depend greatly on the parameter $\mathbf{p}_i$. Therefore it may happen that a limit cycle is stable for a subset P_s of the parameter space P and unstable for its complementary set $P - P_s$.

If we assume that a $(m:n)$ limit cycle is stable for $\mathbf{p}_s \in P$ and unstable for $\mathbf{p}_u \in P$, then there is a continuous curve $\gamma : [-s_0, s_0] \to P$ between the points $\mathbf{p}_s$ and $\mathbf{p}_u$

($\mathbf{p}_s = \gamma(-s_0)$, $\mathbf{p}_u = \gamma(s_0)$). Furthermore, there is at least one eigenvalue $\lambda_i(\gamma(s))$ of the monodromy matrix $D\mathbf{F}(\mathbf{z}_0)$ such that the curve $s \mapsto (\lambda_i \circ \gamma)(s)$ intersects the unit circle K_1 in $\mathbb{C}$ transversally (cf. fig. 3.18). As already mentioned in chapter 2.8.2, three cases have to be distinguished:

(a) The curve $s \mapsto (\lambda_i \circ \gamma)(s)$ intersects K_1 at $(+1, 0)$.

$\Rightarrow$ There might be a pitchfork bifurcation or the transversal intersection induces a turning point, which represents the separatrix between stable and unstable $(m : n)$ cycles.

(b) The curve $s \mapsto (\lambda_i \circ \gamma)(s)$ intersects K_1 at $(-1, 0)$.

$\Rightarrow$ There might be a period doubling from a $(m : n)$ cycle to a $(2m : 2n)$ cycle.

(c) The curve $s \mapsto (\lambda_i \circ \gamma)(s)$ intersects K_1 at $e^{i\vartheta}$ where $\vartheta \neq k\pi$.

$\Rightarrow$ For $\vartheta = \frac{k}{l}\pi$, $k, l \in \mathbb{N}$ a bifurcation to a $(l \cdot m : k \cdot n)$ cycle might occur. For $\vartheta = \kappa \cdot \pi$, $\kappa \in \mathbb{R} - \mathbb{Q}$ a bifurcation to a torus (T^2–HOPF bifurcation) might occur, that is the fix point $\mathbf{z}_0$ bifurcates to a curve in $g^{-1}(0)$ which is homeomorph to S^1 (fig. 3.17).

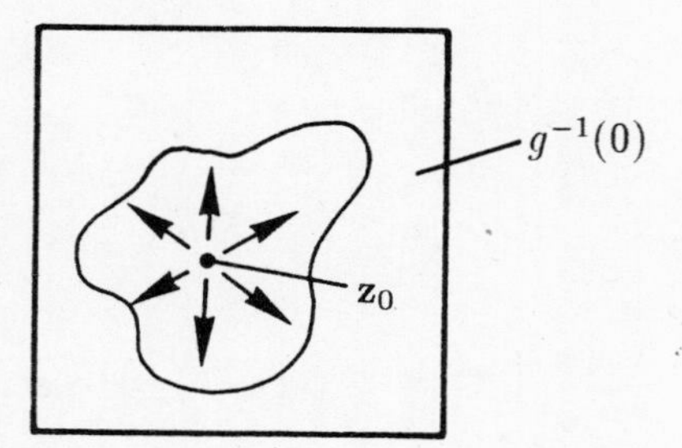

Fig. 3.17: Secundary HOPF bifurcation of a $(m : n)$ cycle

∇

Example 3.7: (one-staged rattling model)

The $(2 : 1)$ limit cycle investigated in example 3.5 depends on the following parameters:

$$p_1 := d \qquad \text{(damping)} \; ,$$

$$p_2 := A \qquad \text{(amplitude of excitation)} \; ,$$

$$p_3 := \omega \qquad \text{(frequency of excitation)} \; ,$$

$$p_4 := \varepsilon \qquad \text{(impact parameter)} \; .$$

In order to investigate the stability of the $(2:1)$ cycle, we need the JACOBIan $D\mathbf{F}(\mathbf{z}_0) \in \mathbb{R}^{2,2}$. $D\mathbf{F}(\mathbf{z}_0)$ has to be computed according to equation (3.25). The determinant of an arbitrary $(m:n)$ cycle is given by

$$det\ (D\mathbf{F}(\mathbf{z}_0)) = \varepsilon^{2m}\, e^{-2d\cdot\left(\frac{2\pi n}{\omega}\right)}\ .$$

By the way, this expression does not depend on the the excitation amplitude A. In case the system is damped $(d > 0)$, we get

$$\lambda_1 \cdot \lambda_2 < 1 \qquad (*)$$

since

$$\varepsilon \leq 1$$

and

$$det\ (D\mathbf{F}(\mathbf{z}_0)) = \lambda_1 \cdot \lambda_2\ .$$

Furthermore, $D\mathbf{F}(\mathbf{z}_0)$ is a real matrix, that is $\lambda_2 = \overline{\lambda}_1$ if λ_1 and λ_2 are complex numbers. Hence, the eigenvalues λ_1 and λ_2 can intersect the unit circle K_1 at ± 1 on the real axis. Otherwise, if the eigenvalues intersect K_1 at $e^{i\vartheta}$ with $\vartheta \neq k\pi$, their modulus $|\lambda_1|$ would be greater than 1 and

$$\lambda_1 \cdot \lambda_2 = |\lambda_1|^2 > 1$$

would apply. This is a contradiction to $(*)$. Therefore, a bifurcation of a $(m:n)$ cycle occurs only at $+1$ or -1 on the real axis (cf. fig. 3.18).

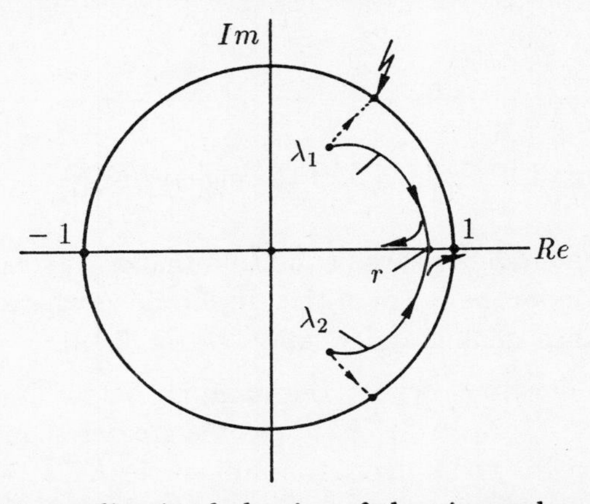

Fig. 3.18: qualitative behavior of the eigenvalues λ_1, λ_2 if the parameters are varied

The eigenvalues merge into the real axis at $r := \lambda_1 = \lambda_2 \in \mathbb{R}$

$$\Rightarrow \qquad r^2 = \varepsilon^{2m}\, e^{-2d\left(\frac{2\pi n}{\omega}\right)} \; .$$

That means the eigenvalues λ_1 and λ_2 meet the real axis at

$$r = \pm\varepsilon^m\, e^{-d\left(\frac{2\pi n}{\omega}\right)} < 1 \; .$$

r is constant in case the amplitude A ($s := A$, $\gamma(s) := (d\,,\, s\,,\, \omega\,,\, \varepsilon)$) is the only varied parameter. Hence, the path $s \mapsto \lambda_1(\gamma(s))$ of eigenvalue λ_1 (and also λ_2) is a concentric circle (radius r) with respect to the unit circle K_1 (cf. fig. 3.19).

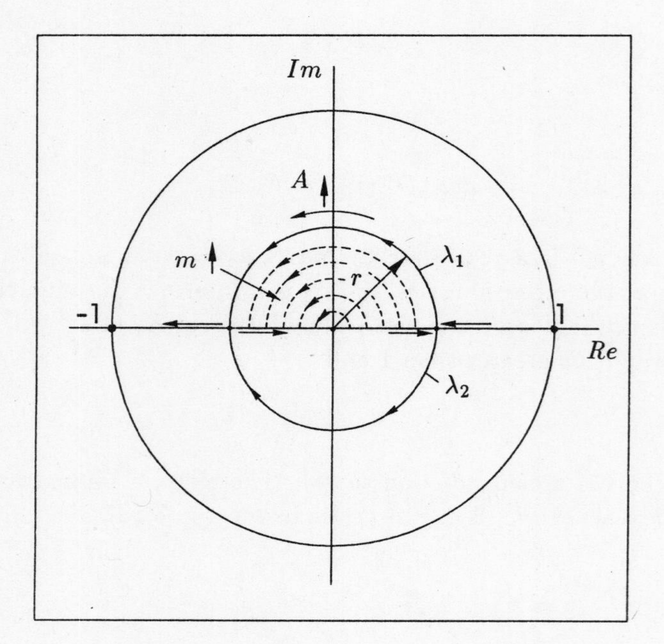

Fig. 3.19: behavior of the eigenvalue $(\lambda_1 \circ \gamma)(A)$

The expression for the radius r says in addition that by increasing the "length" m of the cycle – and therefore the range of the amplitude A where limit cycles are stable – the radius r becomes smaller and smaller (cf. fig. 3.19).

If the parameter d does not denote the damping factor but an excitation factor ($d < 0$), the product $\lambda_1 \cdot \lambda_2 = |\lambda_1|^2$ can become greater than 1 if d is varied. That means the eigenvalues may intersect the unit circle K_1 at $e^{i\vartheta}$, $\vartheta \neq k\pi$ in case of excitation.

Fig. 3.20 shows the behavior of a $(2:1)$ cycle for different parameter values d. The picture on the right hand side shows the phase curve for the relative coordinate x

and the picture on the left hand side represents the phase curve for the coordinate $y(t) = x(t) + e(t)$ with respect to an inertial frame.

A bifurcation for values $d < 0$ might occur if

$$\begin{aligned} 1 &= |\lambda_1|^2 \\ &= \varepsilon^{2m}\, e^{-2d\left(\frac{2\pi n}{\omega}\right)} \end{aligned}$$

holds. That means a HOPF-bifurcation may happen only for the parameter value

$$d^* = \left(\frac{m\omega}{2\pi n}\right) ln\varepsilon < 0 \ .$$

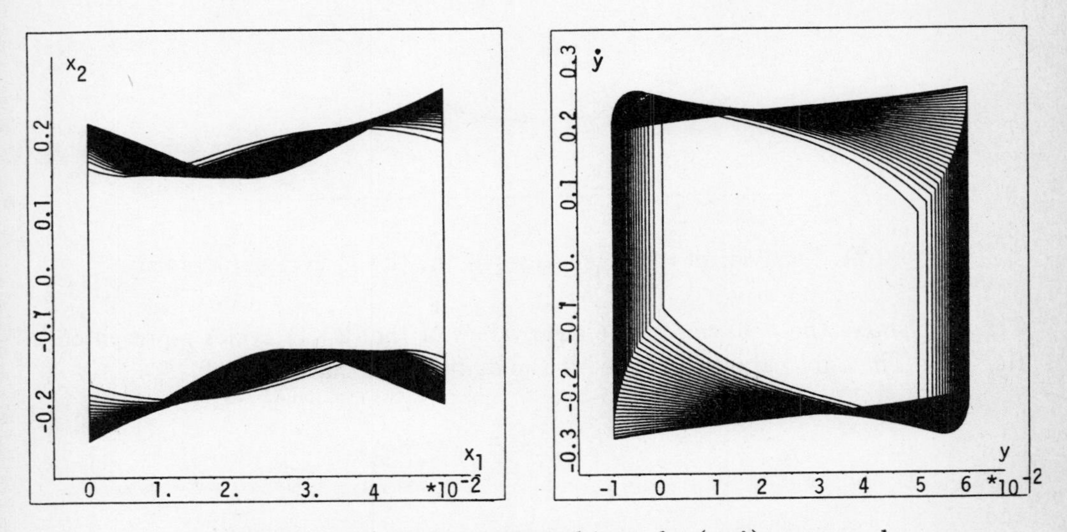

Fig. 3.20: (2 : 1) cycles projected into the $(x, \dot{x})$ space and into the $(y, \dot{y})$ space

The stability domain Z^+ which belongs to the (2 : 1) cycle increases if the cycle length m or the excitation frequency ω is increased or if the impact parameter ε is decreased.

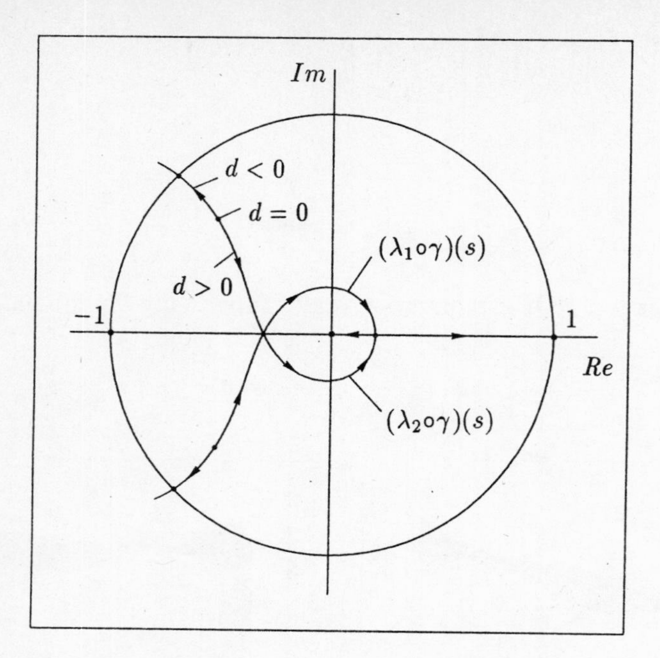

Fig. 3.21: behavior of the eigenvalues of the (2 : 1) cycles if d is varied

Fig. 3.21 shows the behavior of the eigenvalues of the (2 : 1) cycles represented in fig. 3.20. The only parameter varied is the damping d, that is

$$s := d$$

and

$$\gamma(s) := (s\,,\,A\,,\,\omega\,,\,\varepsilon)\ .$$

If $d > 0$, the eigenvalues intersect the imaginary axis from the left hand side to the right hand side in case d is increased.

As mentioned above, there are points $r(d)$ on the real axis where eigenvalues become real, provided the eigenvalues have been complex before (cf. fig. 3.22). While the eigenvalues λ_1 and λ_2 intersect the real axis transversally at $r(d_1)$, they meet at $r(d_2)$ again and stay on the real axis. The limit cycle stays stable as long as one of the eigenvalues intersect the unit circle K_1 at $+1$ what will happen for $d = d_3$. At this point, a bifurcation of the (2 : 1) cycle into three (2 : 1) cycles (one is stable, the other two are unstable) occurs. A further increase of d leads to further bifurcations. Finally the number of limit cycles grows becomes infinite.

A sequence of damping factors $(d_k)_{k\in\mathbb{N}}$ with $\lim_{k\to\infty} d_k = d_g < \infty$ denoting to the bifurcation points spans the bifurcation free. If $\mathbf{z}_0$ is the fix point of any periodic solution, the qualitative behavior of the bifurcation points is shown in fig. 3.22.

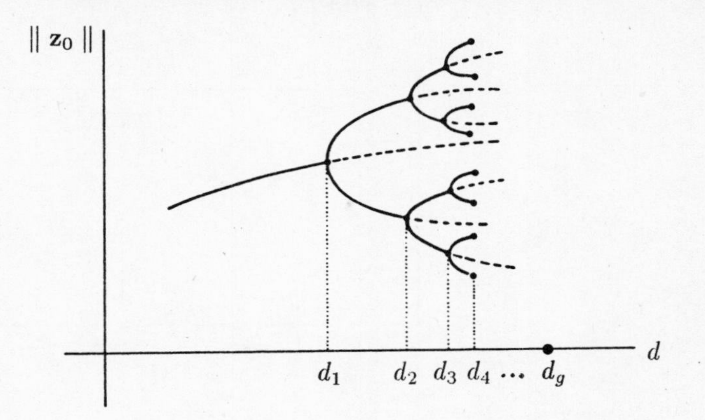

Fig. 3.22: bifurcation tree of (2 : 1) cycles

Once d reaches d_g the system behavior becomes very complex. A further increase of d leads mostly to periodically repeated bifurcation bees. For $d < 0$, the eigenvalues intersect the unit circle at $e^{i\vartheta}$ with $\vartheta \neq k\pi$ if d is decreased. At d^* the $(2:1)$ cycle bifurcates to a $(l\cdot m : k\cdot n)$ cycle $(l, k \in \mathbb{N})$ or to a T^2–torus (T^2–HOPF bifurcation).

△

▽

Example 3.8: (dry friction model)

The dry friction model according to example 3.6 allows to vary the parameters

$$p_1 := \mu \ , \qquad \text{(friction coefficient)} \ ,$$

$$p_2 := \Omega \ , \qquad \text{(eigenfrequency)} \ ,$$

$$p_3 := A \ , \qquad \text{(amplitude of excitation)} \ ,$$

$$p_4 := \omega \ , \qquad \text{(frequency of excitation)}$$

The reference parameter-set $\mathbf{p}_0$ is listed in appendix. The parameters are varied around $\mathbf{p}_0$. In contrary to the rattling model a simple expression for the determinant of the monodromy matrix $DF(\mathbf{z}_0)$ of the dry friction model does not exist. $DF(\mathbf{z}_0)$ has to be computed numerically.

1st case: variation of μ

The reference dataset is $\mathbf{p}_0$. μ is varied between $\mu = 0.05$ and $\mu = 0.9$.

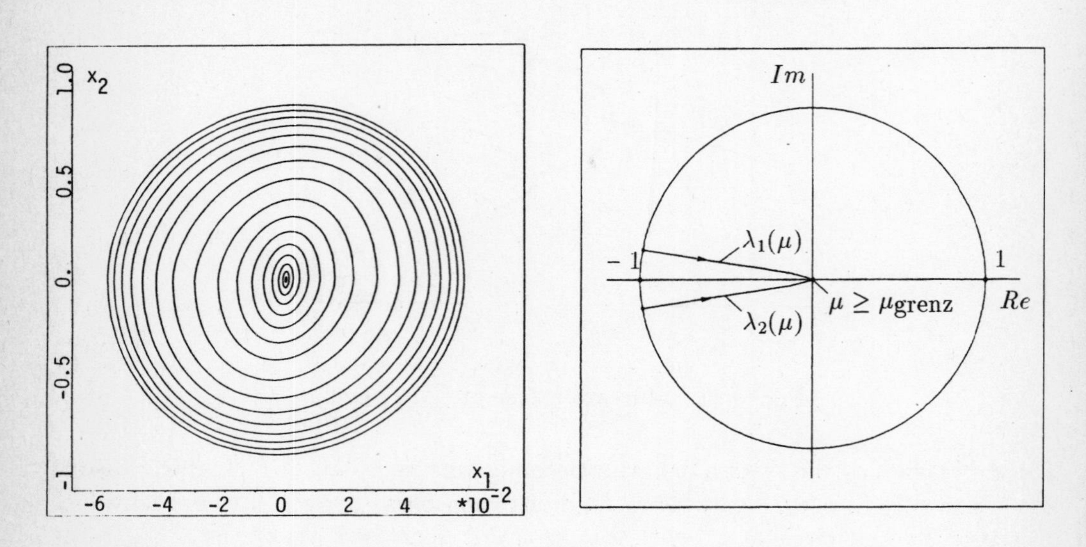

Fig. 3.23: phase curve for the relative coordinates x and $\dot{x}$ and behavior of the eigenvalues for $0.05 \leq \mu \leq 0.95$

Fig. 3.23 shows the phase curves for different values of μ. At $\mu = 0.6$ the limit cycle reaches the sticking area indicated by the kinks at the velocity $v = 0$. The kinks become stronger with increasing μ. As shown in fig. 3.23 the eigenvalues of the monodromy matrix are not influenced during transition into the sticking area. If μ is increased, the eigenvalues move straight into the origin. This phenomenon is reasonable from a physical view point since motion becomes slower and slower if μ is increased. If μ_{limit} indicates the value of μ, where motion stops for any state $\mathbf{z} = (x_1, 0, t)$, we get

$$\Psi(\mathbf{z}) = \mathbf{z} + T\,\mathbf{e}_3 \qquad \forall T \in \mathbb{R}$$

and therefore

$$\mathbf{F}(\mathbf{z}) = T\,\mathbf{e}_3 \ . \quad \Rightarrow \quad D\mathbf{F}(\mathbf{z}) = \mathbf{0} \ .$$

That is, the monodromy matrix for values $\mu > \mu_{limit}$ is the zero matrix, and a bifurcation from a fix point (no motion) to a limit cycle can be considered as a – primary – HOPF bifurcation.

For $\mu \to 0$, the eigenvalues move to the unit circle in $\mathbb{C}$. However a – secondary – HOPF bifurcation may only occur if the excitation frequency ω and the eigenfrequency Ω are in a rational relationship, that is if $\frac{\Omega}{\omega} \in \mathbb{Q}$ holds. If $\frac{\Omega}{\omega} = \frac{m}{n}$ applies, a transversal intersection of the eigenvalues through the unit circle takes place at $e^{2\pi i(\frac{m}{n})}$ (cf. equation (2.173)). If the relationship $\frac{\Omega}{\omega}$ is irrational, the motion becomes ergodic at $\mu = 0$.

2nd case: variation of ω

Fig. 3.24 shows the phase curves and the behavior of the eigenvalues for $10\frac{1}{s} \leq \omega \leq 40\frac{1}{s}$ (reference dataset $\mathbf{p}_0$ presumed).

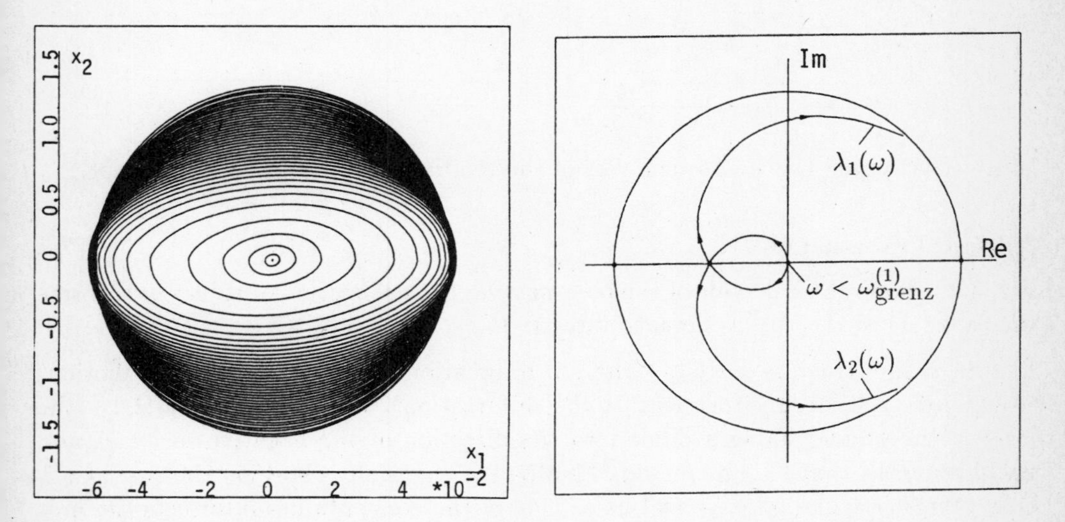

Fig. 3.24: phase curves projected into the $(x, \dot{x})$ plane and behavior of the eigenvalues for $10\frac{1}{s} \leq \omega \leq 40\frac{1}{s}$

If ω is increased, also the amplitude of the limit cycles will increase. For $\mu > 0$ there is a $\omega^{(1)}_{limit}$, where a stable fix point (no motion) bifurcates to a limit cycle (primary HOPF bifurcation).

The limit cycles are stable unless – by increasing ω – the eigenvalues intersect the unit circle transversally (secondary HOPF bifurcation). In contrary to case 1 a T^2–HOPF bifurcation might be possible for $\mu > 0$.

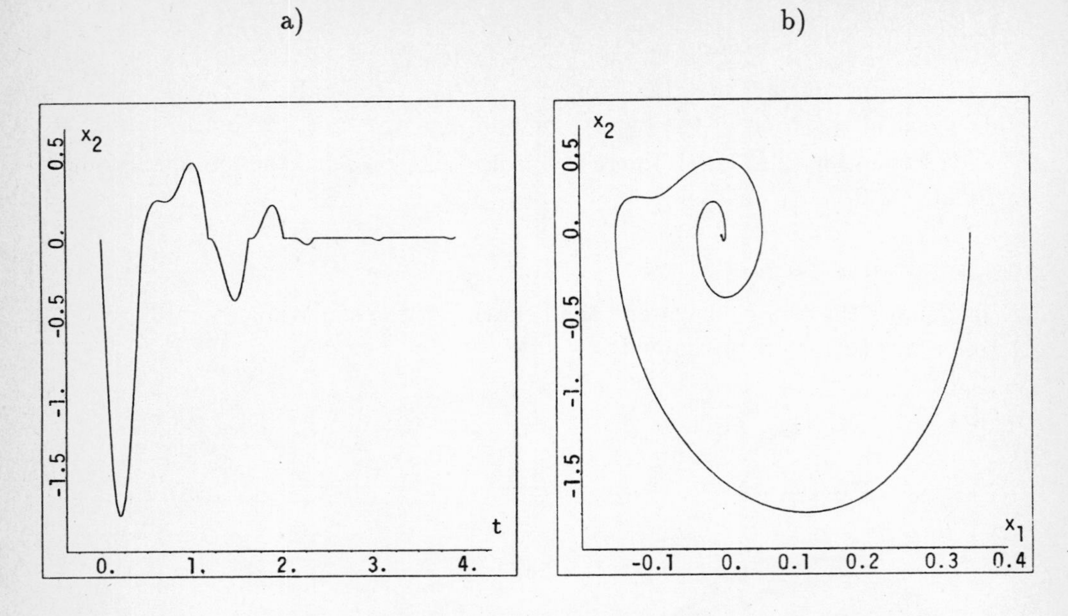

Fig. 3.25: phase curve and behavior of the relative velocity for values $\omega < \omega_{limit}^{(1)}$

<u>3rd case:</u> variation of Ω

Fig. 3.26 shows the behavior of the phase curves for a collection of values of Ω inside the range $9\frac{1}{s} \leq \Omega \leq 40\frac{1}{s}$ (reference value $\mathbf{p_0}$).

Like in case 2 there is a Ω_{limit}, where a bifurcation from a fix point (no motion) to a limit cycle occurs (primary HOPF bifurcation). For values $\Omega > \Omega_{limit}$ the eigenvalues wander along a circle into the direction of the negative real axis and back, provided that Ω is increased. Finally, the eigenvalues stay on the real axis if they merge into the positive real axis. One of these eigenvalues intersects the unit circle at $+1$. At this point a pitchfork bifurcation will occur. The sticking area is left if the eigenvalues intersect the negative real axis.

The results of chapter 2.8.3, obtained from the theory of normal forms according to chapter 2.2, can be applied – without any restriction – to systems with discontinuities. That means the non-linear system is equivalent to the linearized system around a stable limit cycle, provided that a stable limit cycle exists. Loss of stability is equivalent to the resonance of at least one eigenvalue of the monodromy matrix.

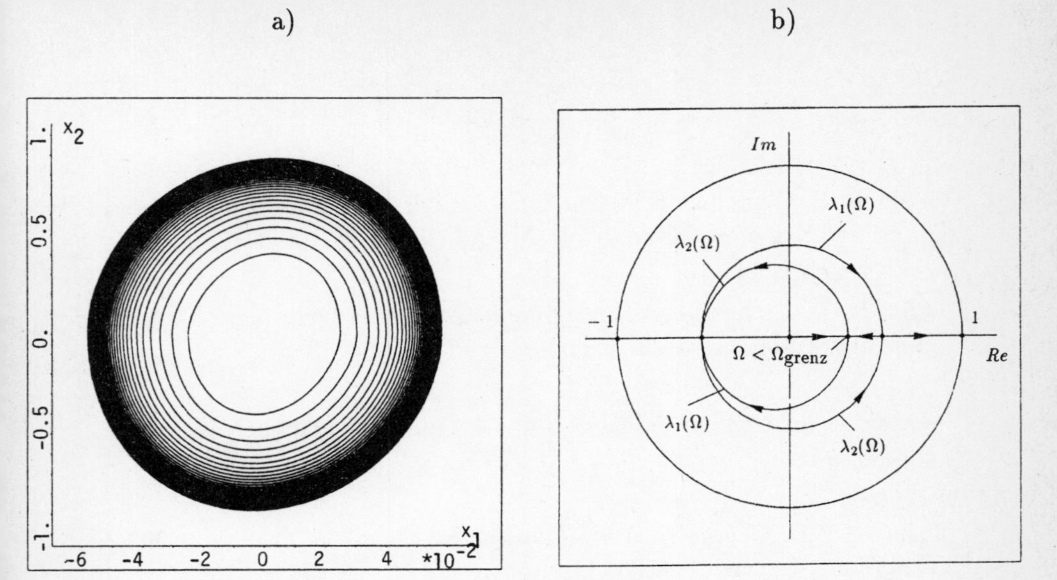

Fig. 3.26: phase curves for the relative coordinate x and behavior of the eigenvalues for $9\frac{1}{s} \leq \Omega \leq 40\frac{1}{s}$

△

4 Literature

[AMANN, ZEHNDER 1980]
Amann, H.; Zehnder, E.: Periodic Solutions of Asymptotically Linear Hamiltonian Systems. Manuscripta Math. 32, 149-189 (1980)

[ANOSOV, ARNOLD 1988]
Anosov, D.V.; Arnold, V.I.: Dynamical Systems I. Springer Verlag, Encyclopedia of Methematical Sciences, vol. 1, 232 p. (1988)

[ARNOLD 1975]
Arnold, V.I.: Critical Points of Smooth Functions and their Normal Forms. Russian Math. Surveys 30:5, 1-75 (1975)

[ARNOLD 1983]
Arnold, V.I.: Geometrical Methods in the Theory of Ordinary Differential Equations. Springer Verlag, Grundlehren 250, XI, 334 p. (1983)

[ARNOLD 1984]
Arnold, V.I.: Mathematical Methods of Classical Mechanics. Springer Verlag, GTM 60, 462 p. (1984)

[ARNOLD, GUSEIN-ZADE, VARCHENKO 1985]
Arnold, V.I.; Gusein-Zade, S.M.; Varchenko, A.N.: Singularities of Differentiable Maps, vol. I. Birkhäuser Verlag, Monographs in Mathem. 82, 382 p. (1985)

[ARNOLD 1988]
Arnold, V.I.: Bifurcations and Singularities in Mathematics and Mechanics. XVIIth International Congress of Theoretical and Applied Mechanics, Aug. 22, (1988)

[AWREJCEWICZ 1989]
Awrejcewicz, J.: An Analytical Method for Detecting HOPF–Bifurcation Solutions in Non–Stationary Non-Linear Systems. J. of Sound and Vibration 129(1), 175-178 (1989)

[BAPAT, POPPLEWELL, McLACHLAN 1983]
Babat, C.N.; Popplewell, N.; McLachlan, K.: Stable Periodic Motions of an Impact-Pair. Journal of Sound and Vibration 87(1), 19-40 (1983)

[BERGER 1970]
Berger, M.S.: Periodic Solutions of Second Order Dynamical Systems and Isoperimetric Variational Problems. American Journal of Mathematics, 1-10, (1970)

[BERGER 1971]
On a Family of Periodic Solutions of Hamiltonian Systems. Journal of Differential Equations 10, 17-26 (1971)

[BIRKHOFF, LEWIS 1933]
Birkhoff, G.D.; Lewis, D.C.: On the Periodic Motions Near a Given Periodic Motion of a Dynamical System. Ann. Mat. Pura Appl. 12, 117-133 (1933)

[BOMMEL 1961]
Bommel van, P.: Berechnung von Zweimassen-Federsystemen mit trockener Reibung für Fahrzeuge. Eisenb.-Techn. Rundschau Archiv 15, pp. 73-88 (1961)

[BOTTKOL 1977]
Bottkol, M.: Bifurcation of Periodic Orbits on Manifolds, and Hamiltonian Systems. Bulletin of the AMS 83, no.5, 1060-1062 (1977)

[BREMER 1988]
Bremer, H.: Dynamik und Regelung mechanischer Systeme. Teubner Studienbücher Mechanik, B.G. Teubner Stuttgart (1988)

[BULIRSCH 1966]
Bulirsch, R.; Stoer, J.: Numerical Treatment of Ordinary Differential Equations by Extrapolation Methods. Num. Math. 8, 1-13 (1966)

[BULIRSCH 1971]
Bulirsch, R.: Die Mehrzielmethode zur numerischen Lösung von nichtlinearen Randwertproblemen und Aufgaben der optimalen Steuerung. Report der CCG (1971)

[BULIRSCH, OETTLI, STOER 1975]
Bulirsch, R.; Oettli, W.; Stoer, J. (ed.): Optimization and Optimal Control. Proceedings of a Conference Held at Oberwolfach, Lecture Notes in Math. vol. 477 (1975)

[BULIRSCH, STOER, DEUFLHARD 1977]
Bulirsch, R.; Stoer, J.; Deuflhard, P.: BOUNDSOL, a Fortrancode for a Numerical Solution of Boundary Value Problems of ODE's. Technische Universität München (1977)

[CAPOZZI, SALVATORE 1982]
Capozzi, A.; Alvatore, A.: Periodic Solutions for Nonlinear Problems with Strong Resonance at Infinity. Comm. Math. Univ. Car. 23, 3, 415-425 (1982)

[CAPOZZI, FORTUNATO, SALVATORE 1985]
Capozzi, A.; Fortunato, D.; Salvatore, A.: Periodical Solutions of Dynamical Systems. Acta Meccania 20, 281-284 (1985)

[CARR 1981]
Carr, J.: Application of Centre Manifold Theory. Springer Verlag, Appl. Math. Sci. 35, 142 p (1981)

[CLARK 1978]
Clark, D.C.: Periodic Solutions of Variational Systems of Ordinary Differential Equations. Journal of Differential Equations 28, 354-368 (1978)

[CLARKE 1979]
Clarke, F.H.: A Classical Variational Principle for Periodic Hamiltonian Trajectories. Proceedings of the AMS vol. 76, no. 1, 186-188 (1979)

[CODDINGTON, LEVINSON 1955]
Coddington, E.A.; Levinson, N.: Theory of Ordinary Differential Equations. McGraw Hill Book Comp., Inc. 429 p. (1955)

[CRONIN 1964]
Cronin, J.: Fixed Points and Topological Degree in Nonlinear Analysis. AMS, Mathematical Surveys no. 11, 198 p. (1964)

[CUSHMAN 1983]
Cushman, R.: Periodic Solutions of the Autonomous Josephson Equation. Springer Verlag, Lecture Notes in Mathematics 985, 319-326 (1983)

[CUSHMAN, DEPRIT, MOSAK 1983]
Cushman, R.; Deprit, A.; Mosak, R.: Normal Form and Representation Theory. J. Math. Phys. 24, 2103-2116 (1983)

[CUSHMAN 1984]
Cushman, R.: Normal Form for Hamiltonian Vector Fields with Periodic Flows. Diff. Geom. Meth. in Math. Physics, ed. S. Sternberg, Reidel, Dordrecht, Holland, 125-144 (1984)

[DEPRIT, HEURARD 1968]
Deprit, A.; Heurard, J.: A Manifold of Periodic Orbits. Adv. Astron. Astroph. 6, 1-124 (1968)

[DEUFLHARD 1984]
Deuflhard, P.: Computation of Periodic Solutions of Nonlinear ODE's. BIT 24, 456-466 (1984)

[DIEKHOFF 1977]
Diekhoff, H.-J.; Lory, P.; Oberle, H.J.; Pesch, H.J.; Rentrop, P.; Seydel, R.: Comparing Routines for the Numerical Solution of Initial Value Problems of ODE's in Multiple Shooting. Num. Math. 27, 449-469 (1977)

[DUISTERMAAT 1981]
Duistermaat, H.: Periodic Solutions Near Equilibrium Points of Hamiltonian Systems. Analytical and Numerical Approaches to Asymptotic Problems in Analysis. Proceed. Conf. Nijmegen, ed. O. Axelsson e.a., North Holland Math. Studies 47, 27-33 (1981)

[DUISTERMAAT 1984]
Duistermaat, H.: Bifurcation of Periodic Solutions Near Equilibrium Points of Hamiltonian Systems. Springer Verlag, Lect. Notes in Math. 1057, 57-105 (1984)

[DULAC 1908]
Dulac, H.: Détermination et intégration d'une certaine classe d'equations

différentielles ayant pour point singulier un centre. Bull. Sci. math.(2) 32, 230-252 (1908)

[EKELAND 1979]
Ekeland, I.: Periodic Solutions of Hamiltonian Equations and a Theorem by P. Rabinowitz. Journal of Differential Equations 34, 523-534 (1979)

[FÜHRER 1988]
Führer, C.: Differential-algebraische Geichungssysteme in mechanischen Mehrkörpersystemen. Theorie, numerische Ansätze und Anwendungen. Dissertationsschrift am Mathematischen Institut der TU München (1988)

[FULLER 1966]
Fuller, F.B.: The Treatment of Periodic Orbits by the Methods of Fixed Point Theory. Bull. of the AMS, vol. 72, 838-840 (1966)

[FULLER 1967]
Fuller, F.B.: An Index of Fixed Point Type for Periodic Orbits. Am. J. Math. 89, 133-148 (1967)

[GORDON 1971]
Gordon, W.B.: A Theorem on the Existence of Periodic Solutions to Hamiltonian Systems with Convex Potential. Journal of Differential Equations 10, 324-335 (1971)

[GRIEPENTROG, MÄRZ 1986]
Griepentrog, E.; März, R.: Differential Algebraic Equations and their Numerical Treatment. B.G. Teubner Verlag (1986)

[GUCKENHEIMER, HOLMES 1983]
Guckenheimer, J.; Holmes, P.: Nonlinear Oscillations, Dynamical Systems, and Bifurcation of Vector Fields. Appl. Math. Sci. 42, Springer Verlag (1983)

[HARRIS 1968]
Harris, T.C.: Periodic Solutions of Arbitrarily Long Periods in Hamiltonian Systems. Journal of Differential Equations 4, 131-141 (1968)

[HARTMAN 1964]
Hartmann, P.: Ordinary Differential Equations. John Wiley & Sons Inc., 612 p. (1964)

[HARTOG 1931]
Hartog, J.P.: Forced Vibrations with Combined Viscous and Coulomb Damping. Trans. ASME, Papers, APM 53-9 (1931)

[HAYASHI 1964]
Hayashi, C.: Nonlinear Oscillations in Physical Systems. Mc Graw Hill Book Comp., 392 p. (1964)

[HEIMANN, BAJAJ, SHERMAN 1988]
Heimann, M.S.; Bajaj, A.K.; Sherman, P.J.: Periodic Motions and Bifur-

cations in Dynamics of an Inclined Impact Pair. Journal of Sound and Vibration 124(1), 55-78 (1988)

[HESTENS 1975]
Hestens, M.R.: Optimization Theory – The Finite Dimensional Case. John Wiley & Sons, 447 p. (1975)

[HOLMES 1982]
Holmes, P.J.: The Dynamics of Repeated Impact With Sinusoidally Vibrating Table. Journal of Sound and Vibration 84, 173-189 (1982)

[HOLODNIOK, KUBIČEK 1984]
Holodniok, M.; Kubiček, M.: DERPER – An Algorithm for the Continuation of Periodic Solutions in Ordinary Differential Equations. Journal of Comput. Physics 55, 254-267 (1984)

[HOPF 1942]
Hopf, E.: Abzweigung einer periodischen Lösung von einer stationären Lösung eines Differentialgleichungssystems. Ber. Math. Phys. Sächsische Akademie der Wissenschaften Leipzig 94, 1-22 (1942)

[HOPF 1955]
Hopf, E.: Repeated Branching Through Loss of Stability, an Example. Proc. Conf. on Diff. Equations, Univ. of Maryland (1955)

[HSU 1977]
Hsu, C.S.: On Nonlinear Parametric Excitation Problems. Advances in Applied Mechanics, vol. 17, 245-301 (1977)

[HSU 1980a]
Hsu, C.S.: Theory of Index for Dynamical Systems of Order Higher Than Two. Journal of Applied Mechanics, vol. 47, 421-427 (1980)

[HSU 1980b]
Hsu, C.S.: Theory of Index and a Generalized NYQUIST Criterion. Int. J. Nonlinear Mechanics, vol. 15, 349-354 (1980)

[JASCHINSKI 1987]
Jaschinski, A.: Anwendung der Kalker'schen Rollreibungstheorie zur dynamischen Simulation von Schienenfahrzeugen. Forschungsbericht der DFVLR, Oberpfaffenhofen, Nr. 87-07 (1987)

[KARAGIANNIS 1989]
Karagiannis, K.: Analyse stoßbehafteter Schwingungssysteme mit Anwendung auf Rasselschwingungen beim Zahnradgetriebe. Dissertation am Lehrstuhl B für Mechanik, TU München (1989)

[KLINGENBERG 1978]
Klingenberg, W.: Lectures on Closed Geodesics. Springer Verlag, Grundlehren 230, 227 p. (1978)

[KRASONSELŠKII, BURD, KOLESOV 1973]
Krasonselškii, M.A.; Burd, V.S.; Kolesov, Y.S.: Nonlinear Almost Periodic Oscillations. John Wiley & Sons, 326 p. (1973)

[LAZER 1975]
Lazer, A.C.: Topolgical Degree and Symmetric Families of Periodic Solutions of Nondissipative Second Order Systems. Journal of Differential Equations 19, 62-69 (1975)

[LEITMANN 1966]
Leitmann, G.: An Introduction to Optimal Control. Mc Graw-Hill Book Comp., New York (1966)

[LEITMANN 1986]
Leitmann, G.: The Calculus of Variations and Optimal Control. Plenum Press, Math. Concepts and Meth. in Sci. and Eng. vol. 24, 311 p. (1986)

[LU 1976]
Lu, Y.C.: Singularity Theory and an Introduction to Catastrophy Theory. Springer Verlag, Universitext, 199 p. (1976)

[LUENBERGER 1969]
Luenberger, D.G.: Optimization by Vector Space Methods. John Wiley & Sons (1969)

[MAGNUS 1955]
Magnus, K.: Über ein Verfahren zur Untersuchung nichtlinearer Schwingungs– und Regelungssysteme. VDI–Forschungsheft 451, Ausgabe B Bd. 21, VDI Verlag Düsseldorf (1955)

[MAGNUS 1976]
Magnus, K.: Schwingungen. Teubner Verlag. Leitfäden der angewandten Math. und Mechanik 251 p. (1976)

[MARKUS 1960]
Markus, L.: The Behavior of the Solutions of a Differential System Near a Periodic Solution. Annals of Mathematics, vol. 72, no. 2, 245-266 (1960)

[MARSDEN, McCRACKEN 1976]
Marsden, J.E.; McCracken, M.: The Hopf Bifurcation and its Applications. Springer Verlag, Appl. Math. Sci. 19, 408 p. (1976)

[MEIJAARD, DE PATER 1989]
Meijaard, J.P.; De Pater, A.D.: Railway Vehicle Systems Dynamics and Chaotic Vibrations. Int. J. Non-Linear Mechanics, vol. 24, no. 1, 1-17 (1989)

[MEYER 1970]
Meyer, K.R.: Generic Bifurcation of Periodic Points. Trans. AMS 149, 95-107 (1970)

[MEYER 1971]
Meyer, K.R.: Generic Stability Properties of Periodic Points. Trans. AMS 154, 273-277 (1971)

[MEYER 1984]
Meyer, K.R.: Normal Forms for the General Equilibrium. Funk. Ekv. 27, 261-271 (1984)

[MICKENS 1988]
Mickens, R.E.: Bounds on the Fourier Coefficients for the Periodic Solutions of Nonlinear Oscillator Equations. Journal of Sound and Vibration 124(1), 199-203 (1988)

[MIL'SHTEIN 1977]
Mil'shtein, G.N.: Stability and Stabilization of Periodic Motions of Autonomous Systems. PMM, vol. 41, no. 4, 744-749 (1977)

[MOSER 1953]
Moser, J.: Periodische Lösungen des restringierten Dreikörperproblems, die sich erst nach vielen Umläufen schließen. Math. Ann. 126, 325-335 (1953)

[MOSER 1976]
Moser, J.: Periodic Orbits Near an Equilibrium and a Theorem by Alan Weinstein. Comm. on Pure and App. Math., vol. 29, 727-747 (1976)

[MÜLLER 1981]
Müller, P.C.: Zur Stabilität von Grenzzyklen. ZAMM 61, T49-T51 (1981)

[NEWHOUSE, RUELLE, TAKENS 1978]
Newhouse, S.; Ruelle, D.; Takens, F.: Occurrence of Strange Axiom - A Attractors Near Quasiperiodic Flow on $T^n, n \leq 3$. Comm. Math. Phys. 64:35 (1978)

[PETZOLD, LÖTSTEDT 1986]
Petzold, L.R.; Lötstedt, P.: Numerical Solution of Nonlinear Differential Equation with Algebraic Constraints. SIAM J. Sci. Stat. Comp. 7, 720-733 (1986)

[PFEIFFER 1984]
Pfeiffer, F.: Mechanische Systeme mit unstetigen Übergängen. Ingenieur Archiv, Bd. 54, Nr. 3, 232-240 (1984)

[PFEIFFER, KÜCÜKAY 1985]
Pfeiffer, F.; Kücücay, F.: Eine erweiterte mechanische Stoßtheorie und ihre Anwendung in der Getriebedynamik. VDI-Z, Bd. 127, Nr. 9, 341-349 (1985)

[PFEIFFER 1988 a]
Pfeiffer, F.: Seltsame Attraktoren in Zahnradgetrieben. Ingenieur Archiv, Bd. 58, 113-125 (1988)

[PFEIFFER 1988 b]
Pfeiffer, F.: Theorie des Getrieberasselns. VDI-Berichte Nr. 697, 45-65 (1988)

[PFEIFFER 1988c]
Pfeiffer, F.: Über unstetige, insbesondere stoßerregte Schwingungen. Z. Flugwiss. Weltraumforsch. 12, 358-367 (1988)

[POINCARÉ 1912]
Poincaré, H.: Sur un théorème de géométrie. Rend. Circ. Math. Palermo, vol. 33, 375-407 (1912)

[POINCARÉ 1893]
Poincaré, H.: Les méthodes nouvelles de la méchanic celéste, Bd. 1 und 2, Paris, (1893)

[RABINOWITZ 1978]
Rabinowitz, P.H.: Periodic Solutions of Hamiltonian Systems. Comm. on Pure and Appl. Math., vol. 31, 157-184 (1978)

[RABINOWITZ 1978]
Rabinowitz, P.H.: Generalized Cohomological Index Theories for Lie Group Actions with an Application to Bifurcation Questions for Hamiltonian Systems. Inventiones Math. 45, 139-174 (1978)

[RABINOWITZ 1979]
Rabinowitz, P.H.: Periodic Solutions of a Hamiltonian System on a Prescribed Energy Surface. Journal of Differential Equations 33, 336-352 (1979)

[RABINOWITZ 1982]
Rabinowitz, P.H.: Periodic Solutions of Hamiltonian Systems: A Survey. SIAM J. Math. Anal., vol. 13, 343-352 (1982)

[REITHMEIER 1984]
Reithmeier, E.: Die numerische Behandlung nichtlinearer Schwingungssysteme mit geschlossener Struktur und Anwendungen. VDI Fortschrittberichte, Reihe 11, Nr. 62 (1984)

[REITHMEIER 1988]
Reithmeier, E.: Periodic Solutions of Nonlinear Dynamical Systems Selected by Filter Criteria. ZAMM 69, 370-372 (1989)

[REITHMEIER 1989]
Reithmeier, E.: Numerical Construction, Bifurcation and Stability of Periodic Solutions of Nonlinear Dynamical Systems with Discontinuities. Proceedings of the Symposium on Nonlinear Dynamics in Engineering Systems. University of Stuttgart (1989)

[ROBINSON 1969]
Robinson, C.: Generic Properties of Conservative Systems. American J. of Math., 562-603 (1969)

[ROELS 1971]
Roels, J.: An Extension to Resonant Cases of Ljapunov's Theorem Concerning the Periodic Solutions Near a Hamiltonian Equilibrium. Journal of Differential Equations 9, 300-324 (1971)

[SEIFERT 1945]
Seifert, H.: Periodische Bewegungen mechanischer Systeme. Mathematische Zeitschrift, Bd. 51, 197-216 (1945)

[SENATOR 1970]
Senator, M.: Existence and Stability of Periodic Motions of a Harmonically Forced Impacting System. Journal of the Acoustical Society of America 47, 1390-1397 (1970)

[SEYDEL 1979]
Seydel, R.: Numerical Computation of Branch Points in Ordinary Differential Equations. Num. Math. 32, 51-68 (1979)

[SEYDEL 1979]
Seydel, R.: Numerical Computation of Branch Points in Nonlinear Equations. Num. Math. 33, 339-352 (1979)

[SEYDEL 1981]
Seydel, R.: Numerical Computation of Periodic Orbits that Bifurcate from Stationary Solutions of Ordinary Differential Equations. Appl. Math. and Comp. 9, 257-271 (1981)

[SEYDEL 1983]
Seydel, R.: Berechnung von periodischen Lösungen bei gewöhnlichen Differentialgleichungen. ZAMM 63, T98-T99 (1983)

[SEYDEL 1988]
Seydel, R.: From Equilibrium to Chaos-Practical Bifurcation and Stability Analysis. Elsevier Sci. Publ. Co., Inc., 367 p. (1988)

[SHAW, HOLMES 1983]
Shaw, S.W.; Holmes, P.J.: A Periodically Forced Piecewise Linear Oscillator. Journal of Sound and Vibration 90, 129-155 (1983)

[SHAW 1985]
Shaw, S.W.: The Dynamics of Harmonically excited System Having Rigid Constraints, Part 1. J. App. Mech. 52, 453-464 (1985)

[SIEGEL 1954]
Siegel, C.L.: Über die Existenz einer Normalform analytischer Hamilton'scher Differentialgleichungen in der Nähe einer Gleichgewichtslösung. Math. Ann. 128, 144-170 (1954)

[SIEGEL 1956]
Siegel, C.L.: Vorlesungen über Himmelsmechanik. Springer Verlag, Grundlehren 85, 212 p. (1956)

[SIEGEL 1971]
Siegel, C.L.: Periodische Lösungen von Differentialgleichungen. Nachr. Akad. Wiss. Göttingen. Math. Phys. Kl. II, 261-283 (1971)

[SIMEON 1988]
Simeon, B.: Homotopieverfahren zur Berechnung quasistationärer Lagen

von Deskriptorformen in der Mechanik. Diplomarbeit am Institut für Mathematik der TU München (1988)

[STEIDEL 1989]
Steidel, R.F. Jr.: An Introduction to Mechanical Vibrations. John Wiley & Sons, New York (1989)

[STOER, BULIRSCH 1979]
Stoer, J.; Bulirsch, R.: Introduction to Numerical Analysis. Springer Verlag (1979)

[TAKENS 1974]
Takens, F.: Singularities of Vector Fields. Publ. Math. IHES 43, 47-100 (1974)

[THOMPSON, STEWART 1986]
Thompson, J.M.T.; Stewart, H.B.: Nonlinear Dynamics and Chaos. John Wiley and Sons. Chichester - New York - Brisbane - Toronto - Singapore (1986)

[TOUSI, BAJAJ 1985]
Tousi, S.; Bajaj, A.K.: Period-Doubling Bifurcations and Modulated Motions in Forced Mechanical Systems. Trans. ASME, vol. 52, 446-452 (1985)

[TROGER, ZEMAN 1981]
Troger, H.; Zeman, K.: Zur korrekten Modellbildung in der Dynamik diskreter Systeme. Ing. Archiv 51, 31-43 (1981)

[TROGER, ZEMAN 1981]
Troger, H.; Zeman, K.: Application of Bifurcation Theory to Tractor–Semitrailer Dynamics. Proc. of the 7-th Symp. on Vehicle Dynamics. Cambridge 156-160 (1981)

[URABE 1967]
Urabe, M.: Nonlinear Autonomous Oscillations. Academic Press, Math. in Sci. and Eng. vol. 34, 330 p. (1967)

[WEINSTEIN 1973]
Weinstein, A.: Normal Modes for Nonlinear Hamiltonian Systems. Inv. Math. 20, 47-57 (1973)

[WEINSTEIN 1978]
Weinstein, A.: Periodic Orbits for Convex Hamiltonian Systems. Ann. Math. 108, 507-518 (1978)

[WEINSTEIN 1978]
Weinstein, A.: Bifurcation and Hamilton's Principle. Math. Z. 159, 235-248 (1978)

[WEINSTEIN 1979]
Weinstein, A.: On the Hypothesis of Rabinowitz' Periodic Orbit Theorems. Journal of Differential Equations 33, 353-358 (1979)

[YOSHIZAWA 1975]
Yoshizawa, T.: Stability Theory and the Existence of Periodic Solutions and Almost Periodic Solutions. Appl. Math. Sci 14, Springer Verlag, 233p (1975)

5 Appendix – dataset

- Example 2.4:

$$\begin{aligned} \alpha &= 0.1 \ , \\ \beta &= 1.0 \ , \\ \varepsilon &= 5.0 \cdot 10^{-3} \ , \\ \gamma &= 1.0 \ , \\ \omega &= 2\pi \ . \end{aligned}$$

- Example 2.7: see example 2.4 !

- Example 2.14:

$$\begin{aligned} m_1 &= m_2 = 1\,kg \ , \\ l_1 &= l_2 = 0.1\,m \ , \\ g &= 9.81\,\tfrac{m}{s^2} \ . \end{aligned}$$

- Chapter 2.6.2: (according to [JASCHINSKI 1987])

  - geometry:

$$\begin{aligned} a &= 0.3\,m \ , \\ r_0 &= 0.1\,m \ , \\ \delta &= 0.02\,rad \ . \end{aligned}$$

  - masses, masses of inertia:

$$\begin{aligned} m &= 16\,kg \ , \\ I_x &= 0.06\,kg\,m^2 \ , \\ I_y &= 0.36\,kg\,m^2 \ , \\ I_z &= 0.36\,kg\,m^2 \ . \end{aligned}$$

  - slip model:

$$\begin{aligned} \mu &= 0.3 \ , \\ G &= 7.9 \cdot 10^{10}\,\tfrac{N}{mm^2} \ , \\ k_x &= C_{11} \cdot k^* \ , \\ k_y &= C_{22} \cdot k^* \ , \\ k_z &= C_{23} \cdot b \cdot k^* \ , \\ k^* &= G \cdot b^2 \,/\, \mu \ , \\ b &= 1.05 \cdot m \cdot g \cdot e \,/\, \left[2\pi \cdot (d+c) \cdot G \cdot \sqrt{g_e}\right] \\ d &= \cos\delta \,/\, 2\,r_0 \ , \\ c &= 1\,/\,2 \cdot r' \ , \\ g_e &= 0.71 \ , \\ e &= 1.35 \ , \\ r' &= 0.06\,m \ , \\ C_{11} &= 4.72 \ , \\ C_{22} &= 4.27 \ , \\ C_{23} &= 4.97 \ . \end{aligned}$$

$$
\begin{aligned}
c_x &= 5.0 \cdot 10^5 \, \tfrac{N}{m} \\
c_y &= 5.0 \cdot 10^5 \, \tfrac{N}{m} \\
c_z &= 5.0 \cdot 10^5 \, \tfrac{N}{m} \\
c_\gamma &= 4.5 \cdot 10^3 \, \tfrac{Nm}{rad} \\
c_\alpha &= 4.5 \cdot 10^3 \, \tfrac{Nm}{rad}
\end{aligned}
$$

- Example 2.19: see example 2.14 !
- Example 2.21: see chapter 2.6.2 !
- Example 2.22: see chapter 2.6.2 !
- Example 2.23: see example 2.14 !
- Example 3.5: (reference dataset)

$$
\begin{aligned}
\omega &= 10.0 \, \tfrac{1}{s} \; , \\
\delta &= 2.3 \, \tfrac{Ns}{m} \; , \\
\varepsilon &= 0.8 \; , \\
a &= 0.01 \, m \; , \\
v &= 0.05 \, m \; , \\
m &= 1.0 \, kg \; .
\end{aligned}
$$

- Example 3.6: (reference dataset)

$$
\begin{aligned}
\omega &= 10.0 \, \tfrac{1}{s} \; , \\
\mu &= 0.35 \; , \\
a &= 0.05 \, m \; , \\
m &= 1.0 \, kg \; , \\
c &= 50 \, \tfrac{N}{m} \; .
\end{aligned}
$$

- Example 3.7: see example 3.5 !
- Example 3.8: see example 3.6 !

6 Index

Vol. 1394: T.L. Gill, W.W. Zachary (Eds.), Nonlinear Semigroups, Partial Differential Equations and Attractors. Proceedings, 1987. IX, 233 pages. 1989.

Vol. 1395: K. Alladi (Ed.), Number Theory, Madras 1987. Proceedings. VII, 234 pages. 1989.

Vol. 1396: L. Accardi, W. von Waldenfels (Eds.), Quantum Probability and Applications IV. Proceedings, 1987. VI, 355 pages. 1989.

Vol. 1397: P.R. Turner (Ed.), Numerical Analysis and Parallel Processing. Seminar, 1987. VI, 264 pages. 1989.

Vol. 1398: A.C. Kim, B.H. Neumann (Eds.), Groups – Korea 1988. Proceedings. V, 189 pages. 1989.

Vol. 1399: W.-P. Barth, H. Lange (Eds.), Arithmetic of Complex Manifolds. Proceedings, 1988. V, 171 pages. 1989.

Vol. 1400: U. Jannsen. Mixed Motives and Algebraic K-Theory. XIII, 246 pages. 1990.

Vol. 1401: J. Steprans, S. Watson (Eds.), Set Theory and its Applications. Proceedings, 1987. V, 227 pages. 1989.

Vol. 1402: C. Carasso, P. Charrier, B. Hanouzet, J.-L. Joly (Eds.), Nonlinear Hyperbolic Problems. Proceedings, 1988. V, 249 pages. 1989.

Vol. 1403: B. Simeone (Ed.), Combinatorial Optimization. Seminar, 1986. V, 314 pages. 1989.

Vol. 1404: M.-P. Malliavin (Ed.), Séminaire d´AIgèbre Paul Dubreil et Marie-Paul Malliavin. Proceedings, 1987–1988. IV, 410 pages. 1989.

Vol. 1405: S. Dolecki (Ed.), Optimization. Proceedings, 1988. V, 223 pages. 1989. Vol. 1406: L. Jacobsen (Ed.), Analytic Theory of Continued Fractions III. Proceedings, 1988. VI, 142 pages. 1989.

Vol. 1407: W. Pohlers, Proof Theory. VI, 213 pages. 1989.

Vol. 1408: W. Lück, Transformation Groups and Algebraic K-Theory. XII, 443 pages. 1989.

Vol. 1409: E. Hairer, Ch. Lubich, M. Roche. The Numerical Solution of Differential-Algebraic Systems by Runge-Kutta Methods. VII, 139 pages. 1989.

Vol. 1410: F.J. Carreras, O. Gil-Medrano, A.M. Naveira (Eds.), Differential Geometry. Proceedings, 1988. V, 308 pages. 1989.

Vol. 1411: B. Jiang (Ed.), Topological Fixed Point Theory and Applications. Proceedings. 1988. VI, 203 pages. 1989.

Vol. 1412: V.V. Kalashnikov, V.M. Zolotarev (Eds.), Stability Problems for Stochastic Models. Proceedings, 1987. X, 380 pages. 1989.

Vol. 1413: S. Wright, Uniqueness of the Injective III_1 Factor. III, 108 pages. 1989.

Vol. 1414: E. Ramirez de Arellano (Ed.), Algebraic Geometry and Complex Analysis. Proceedings, 1987. VI, 180 pages. 1989.

Vol. 1415: M. Langevin, M. Waldschmidt (Eds.), Cinquante Ans de Polynômes. Fifty Years of Polynomials. Proceedings, 1988. IX, 235 pages.1990.

Vol. 1416: C. Albert (Ed.), Géométrie Symplectique et Mécanique. Proceedings, 1988. V, 289 pages. 1990.

Vol. 1417: A.J. Sommese, A. Biancofiore, E.L. Livorni (Eds.), Algebraic Geometry. Proceedings, 1988. V, 320 pages. 1990.

Vol. 1418: M. Mimura (Ed.), Homotopy Theory and Related Topics. Proceedings, 1988. V, 241 pages. 1990.

Vol. 1419: P.S. Bullen, P.Y. Lee, J.L. Mawhin, P. Muldowney, W.F. Pfeffer (Eds.), New Integrals. Proceedings, 1988. V, 202 pages. 1990.

Vol. 1420: M. Galbiati, A. Tognoli (Eds.), Real Analytic Geometry. Proceedings, 1988. IV, 366 pages. 1990.

Vol. 1421: H.A. Biagioni, A Nonlinear Theory of Generalized Functions, XII, 214 pages. 1990.

Vol. 1422: V. Villani (Ed.), Complex Geometry and Analysis. Proceedings, 1988. V, 109 pages. 1990.

Vol. 1423: S.O. Kochman, Stable Homotopy Groups of Spheres: A Computer-Assisted Approach. VIII, 330 pages. 1990.

Vol. 1424: F.E. Burstall, J.H. Rawnsley, Twistor Theory for Riemannian Symmetric Spaces. III, 112 pages. 1990.

Vol. 1425: R.A. Piccinini (Ed.), Groups of Self-Equivalences and Related Topics. Proceedings, 1988. V, 214 pages. 1990.

Vol. 1426: J. Azéma, P.A. Meyer, M. Yor (Eds.), Séminaire de Probabilités XXIV, 1988/89. V, 490 pages. 1990.

Vol. 1427: A. Ancona, D. Geman, N. Ikeda, École d'Eté de Probabilités de Saint Flour XVIII, 1988. Ed.: P.L. Hennequin. VII, 330 pages. 1990.

Vol. 1428: K. Erdmann, Blocks of Tame Representation Type and Related Algebras. XV. 312 pages. 1990.

Vol. 1429: S. Homer, A. Nerode, R.A. Platek, G.E. Sacks, A. Scedrov, Logic and Computer Science. Seminar, 1988. Editor: P. Odifreddi. V, 162 pages. 1990.

Vol. 1430: W. Bruns, A. Simis (Eds.), Commutative Algebra. Proceedings. 1988. V, 160 pages. 1990.

Vol. 1431: J.G. Heywood, K. Masuda, R. Rautmann, V.A. Solonnikov (Eds.), The Navier-Stokes Equations – Theory and Numerical Methods. Proceedings, 1988. VII, 238 pages. 1990.

Vol. 1432: K. Ambos-Spies, G.H. Müller, G.E. Sacks (Eds.), Recursion Theory Week. Proceedings, 1989. VI, 393 pages. 1990.

Vol. 1433: S. Lang, W. Cherry, Topics in Nevanlinna Theory. II, 174 pages.1990.

Vol. 1434: K. Nagasaka, E. Fouvry (Eds.), Analytic Number Theory. Proceedings, 1988. VI, 218 pages. 1990.

Vol. 1435: St. Ruscheweyh, E.B. Saff, L.C. Salinas, R.S. Varga (Eds.), Computational Methods and Function Theory. Proceedings, 1989. VI, 211 pages. 1990.

Vol. 1436: S. Xambó-Descamps (Ed.), Enumerative Geometry. Proceedings, 1987. V, 303 pages. 1990.

Vol. 1437: H. Inassaridze (Ed.), K-theory and Homological Algebra. Seminar, 1987–88. V, 313 pages. 1990.

Vol. 1438: P.G. Lemarié (Ed.) Les Ondelettes en 1989. Seminar. IV, 212 pages. 1990.

Vol. 1439: E. Bujalance, J.J. Etayo, J.M. Gamboa, G. Gromadzki. Automorphism Groups of Compact Bordered Klein Surfaces: A Combinatorial Approach. XIII, 201 pages. 1990.

Vol. 1440: P. Latiolais (Ed.), Topology and Combinatorial Groups Theory. Seminar, 1985–1988. VI, 207 pages. 1990.

Vol. 1441: M. Coornaert, T. Delzant, A. Papadopoulos. Géométrie et théorie des groupes. X, 165 pages. 1990.

Vol. 1442: L. Accardi, M. von Waldenfels (Eds.), Quantum Probability and Applications V. Proceedings, 1988. VI, 413 pages. 1990.

Vol. 1443: K.H. Dovermann, R. Schultz, Equivariant Surgery Theories and Their Periodicity Properties. VI, 227 pages. 1990.

Vol. 1444: H. Korezlioglu, A.S. Ustunel (Eds.), Stochastic Analysis and Related Topics VI. Proceedings, 1988. V, 268 pages. 1990.

Vol. 1445: F. Schulz, Regularity Theory for Quasilinear Elliptic Systems and – Monge Ampère Equations in Two Dimensions. XV, 123 pages. 1990.

Vol. 1446: Methods of Nonconvex Analysis. Seminar, 1989. Editor: A. Cellina. V, 206 pages. 1990.

Vol. 1447: J.-G. Labesse, J. Schwermer (Eds), Cohomology of Arithmetic Groups and Automorphic Forms. Proceedings, 1989. V, 358 pages. 1990.

Vol. 1448: S.K. Jain, S.R. López-Permouth (Eds.), Non-Commutative Ring Theory. Proceedings, 1989. V, 166 pages. 1990.

Vol. 1449: W. Odyniec, G. Lewicki, Minimal Projections in Banach Spaces. VIII, 168 pages. 1990.

Vol. 1450: H. Fujita, T. Ikebe, S.T. Kuroda (Eds.), Functional-Analytic Methods for Partial Differential Equations. Proceedings, 1989. VII, 252 pages. 1990.

Vol. 1451: L. Alvarez-Gaumé, E. Arbarello, C. De Concini, N.J. Hitchin, Global Geometry and Mathematical Physics. Montecatini Terme 1988. Seminar. Editors: M. Francaviglia, F. Gherardelli. IX, 197 pages. 1990.

Vol. 1452: E. Hlawka, R.F. Tichy (Eds.), Number-Theoretic Analysis. Seminar, 1988–89. V, 220 pages. 1990.

Vol. 1453: Yu.G. Borisovich, Yu.E. Gliklikh (Eds.), Global Analysis – Studies and Applications IV. V, 320 pages. 1990.

Vol. 1454: F. Baldassari, S. Bosch, B. Dwork (Eds.), p-adic Analysis. Proceedings, 1989. V, 382 pages. 1990.

Vol. 1455: J.-P. Françoise, R. Roussarie (Eds.), Bifurcations of Planar Vector Fields. Proceedings, 1989. VI, 396 pages. 1990.

Vol. 1456: L.G. Kovács (Ed.), Groups – Canberra 1989. Proceedings. XII, 198 pages. 1990.

Vol. 1457: O. Axelsson, L.Yu. Kolotilina (Eds.), Preconditioned Conjugate Gradient Methods. Proceedings, 1989. V, 196 pages. 1990.

Vol. 1458: R. Schaaf, Global Solution Branches of Two Point Boundary Value Problems. XIX, 141 pages. 1990.

Vol. 1459: D. Tiba, Optimal Control of Nonsmooth Distributed Parameter Systems. VII, 159 pages. 1990.

Vol. 1460: G. Toscani, V. Boffi, S. Rionero (Eds.), Mathematical Aspects of Fluid Plasma Dynamics. Proceedings, 1988. V, 221 pages. 1991.

Vol. 1461: R. Gorenflo, S. Vessella, Abel Integral Equations. VII, 215 pages. 1991.

Vol. 1462: D. Mond, J. Montaldi (Eds.), Singularity Theory and its Applications. Warwick 1989, Part I. VIII, 405 pages. 1991.

Vol. 1463: R. Roberts, I. Stewart (Eds.), Singularity Theory and its Applications. Warwick 1989, Part II. VIII, 322 pages. 1991.

Vol. 1464: D. L. Burkholder, E. Pardoux, A. Sznitman, Ecole d'Eté de Probabilités de Saint- Flour XIX-1989. Editor: P. L. Hennequin. VI, 256 pages. 1991.

Vol. 1465: G. David, Wavelets and Singular Integrals on Curves and Surfaces. X, 107 pages. 1991.

Vol. 1466: W. Banaszczyk, Additive Subgroups of Topological Vector Spaces. VII, 178 pages. 1991.

Vol. 1467: W. M. Schmidt, Diophantine Approximations and Diophantine Equations. VIII, 217 pages. 1991.

Vol. 1468: J. Noguchi, T. Ohsawa (Eds.), Prospects in Complex Geometry. Proceedings, 1989. VII, 421 pages. 1991.

Vol. 1469: J. Lindenstrauss, V. D. Milman (Eds.), Geometric Aspects of Functional Analysis. Seminar 1989-90. XI, 191 pages. 1991.

Vol. 1470: E. Odell, H. Rosenthal (Eds.), Functional Analysis. Proceedings, 1987-89. VII, 199 pages. 1991.

Vol. 1471: A. A. Panchishkin, Non-Archimedean L-Functions of Siegel and Hilbert Modular Forms. VII, 157 pages. 1991.

Vol. 1472: T. T. Nielsen, Bose Algebras: The Complex and Real Wave Representations. V, 132 pages. 1991.

Vol. 1473: Y. Hino, S. Murakami, T. Naito, Functional Dif rential Equations with Infinite Delay. X, 317 pages. 1991.

Vol. 1474: S. Jackowski, B. Oliver, K. Pawałowski (Ed Algebraic Topology, Poznań 1989. Proceedings. VIII, 397 pag 1991.

Vol. 1475: S. Busenberg, M. Martelli (Eds.), Delay Different Equations and Dynamical Systems. Proceedings, 1990. VIII, 2 pages. 1991.

Vol. 1476: M. Bekkali, Topics in Set Theory. VII, 120 pag 1991.

Vol. 1477: R. Jajte, Strong Limit Theorems in Noncommutati L_2-Spaces. X, 113 pages. 1991.

Vol. 1478: M.-P. Malliavin (Ed.), Topics in Invariant Theo Seminar 1989-1990. VI, 272 pages. 1991.

Vol. 1479: S. Bloch, I. Dolgachev, W. Fulton (Eds.), Algebr Geometry. Proceedings, 1989. VII, 300 pages. 1991.

Vol. 1482: J. Chabrowski, The Dirichlet Problem with L Boundary Data for Elliptic Linear Equations. VI, 173 pag 1991.

Vol. 1483: E. Reithmeier, Periodic Solutions of Nonline Dynamical Systems. Vi, 171 pages. 1991.